JUST SKIP DINNER

Fruit and Fat-Fasting Miracles

Written and Illustrated by

Karen Kellock Ph.D.

On the struggles of genius in overcoming obstruction.
Your time will come if you faint not.

FORMULA:

All success attraction
All disease obstruction
All recovery elimination

For Independence & Power, Fast on all three
OBSTRUCTIONS:

People
Habit
Food

JUST SKIP DINNER

Must fast on all obstructions:

People, Habit and Food

Purpose: To reveal the obstructions to your
breakthrough and show how you've been lied to.
The Champion Guides/Manual for Superior Men Series
Deals with the Struggles of Genius in Overcoming
Obstruction. Your Time Will Come If You Faint Not.
Only in Purity Can You See God. To Purify, Turn
From All Bad Associations and Habits of Slobs.

PREFACE

Karen Kellock:

*Recovery through Solitude and Daily
Fasting on People, Habit and Food.*

This theory is based on the work of (and reversal dieting between) Arnold Ehret (fruitarianism) and Robert Atkins (fat-fasting)—both of whom died not through diet but head injuries from slipping on the street. Reversal dieting brings stability: with even emotions and happy moods you can just enjoy your day.

FASTING

"Bliss" is not the Riviera or Vegas but your own home in the fasting state, in which a miraculous new life opens through an inner journey.

Daily Fasting is the Key.

Higher Paleo Fasting sees *fats as essential,* with reversals into short fruit-fasts (like the grapecure) as key. The higher paleo-faster (lacto-fruitarian) eats fruits, cheese, nuts and occasional fish. Now you'll have more beauty than you've ever seen *despite* your genes.

Salve: **A theory that comforts, soothes and solves.**

Take fruit or fat then fast 16 hours, releasing all powers. Or
eat mouse meals (continuous fasting punctuated by bites).
Thus return to selfhood and success (your rights). If fasting,
Human Growth Hormone is released at night. It youthifies,
repairs and fills you with might.

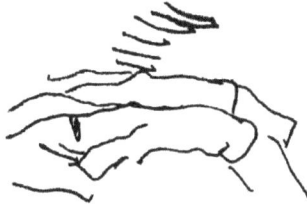

THE DAILY FAST

Written by a fruitarian who found daily fasting after
fruit/fat to be superior and the inner life far superior to all.
Man needs fats: with gut-dense foods like a little cheese with
a fruit diet it requires far less to perfectly sustain. The
fruitarian is someone who eats mostly fruits, but fats burn
fat and create energy with full   appetite suppression so
bountiful health is restored/maintained and the Daily Fast is
enjoyed with all tastes tamed.

11

Daily fasting after fruit and fat is fun--you'll no longer walk you'll run. You'll feel light and free like removing a ton and as the body cleans it beautifies in the sun.

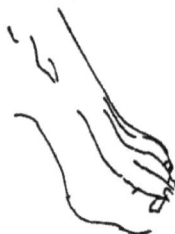

Question: What do you say about Atkins Dieters or Ehretists?

KK: The Atkin's Dieters eat too much, especially too much meat. They eat too little fat (despite how Atkins acclaimed it as essential and good), they eat too many meals, they don't fast enough, they eat too little fruit in fear of carbs and I doubt they'd ever eat raisins or figs-- a "concentrated fruit sugar". The Ehretists in contrast eat too much fruit, don't fast enough, don't eat enough fat (fat-phobic), have mood swings (sugar rushes or detox) and become increasingly dogmatic as the restricted diet fails them ("failure to thrive").

FRUIT-FAT-FAST:
The Heavenly Triangle

*It's so simple—just three is the
superior diet for thee.*

GLOSSARY

<u>Fauna</u>: animal protein and fat

ENANTIODROMIA:

Everything converting to it's opposite: from prison to
palace, from fat to thin. Genius: a dormant potential in all
humans, evoked through released obstruction

<u>Enantiodromia</u> is the inversion of systems whereby the
sinner/tyrant brings himself down just as the champion
rises up. Through fasting on obstructions the top
becomes the bottom and the bottom the top (no more
the wet mop). Now ends the pain caused by
troublemakers.

This book is fun so you can relax while absorbing the facts. Your lack of breakthrough is nothing you lack but an obstruction to remove in fact. Then your genius will flow (you'll have rare tact)

If you've a destiny—made to make a mark— you'll misfit all groups. So in the preparatory stage coming up to success, you're likely to feel a mess (as if you're less). But don't feel stress: God wants you alone before taking the throne.

FAT

elevates glucagons for fat-burning

dilates airways

increases oxygen

vaso-dilates

enhances immunity

decreases pain

decreases inflammation

burns cholesterol

increases endurance

prevents platelet aggregation (heart)

decreases cell proliferation (cancer)

moistens skin

eliminates all water/salt from the kidneys

pathway to good eicosanoids

SUGAR/STARCH

elevates insulin—lays on fat

constricts airways (asthma)

decreases oxygen flow

vaso-constricts

suppresses immunity

increases inflammation

cues liver to create cholesterol

increases pain

decreases endurance

platelet aggregation (heart)

increases cell proliferation (cancer)

dries skin (rash eczema)

retains water and salt (puffiness)

pathway to bad eicosanoids

1
FAST MIRACLES

Your miracle may be finding a mate, being financially great or discovering a gimmick that is first-rate. Fast to be happy, so high as self-esteem shoots to the sky. Food-addicts (the blue): won't you give this a try? To all your problems you'll say good-bye as you seek the true and shun the lie.

◈DISCOVERY!◈

As part of nature the Creative Act evolves through cycles which cannot be rushed. But for those with the courage to work and patience to wait, the entire Creative Act together with the reward itself will fall together in a fabric. It is the FAST which provides this link to success. Amidst chaos the superior man perfects himself and fulfills his destiny—to make the amazing new dent for which he's been sent.

Fasting levels the playing field and negates the need to study and read—it's simple and that's our creed--as it erases all barriers such as age and gender. Do you worry about aging, bemoan past mistakes or suffer the effects of sexism? No more for when you fast you become the possessor of royal privileges and there *are* no more barriers. Just start and you will see the real person you were *meant* to be--no more belittled as a cultural stereotype but valued as a *universal archetype.* The fast gives you greatest dignity making you divine royalty while the actual king who eats incessantly is a fool and a pauper. See the light, become true might. To your delight it excites, it ignites. Just eat tonight but today I invite you to re-unite with God, be His Knight. It's like a flight leaving the blight straight for the bright. It's out of sight becoming a sprite so is the fast a fright? Not by a long sight.

❧ THE HERD SIDES AGAINST THE GREAT ❧
So the Great Must Fast

The faster is an exemplar, a champion standing out in a sea of sharks. Not-eating he's a brilliant, radiant life reflecting the Creator's blueprint—his True Self. He is not led by puny, finite, darkened minds of men but inspired by God's force transcending all culture and silly human fashions. The Champion has the capacity to shift between disparate realities—the good and the bad: he never gets stuck. When others get their claws into him he just goes invisible for in his own inner world nothing can touch him. The herd sides against the Great but fasting overcomes them all--so the Great must fast.

❧ SICK OF THE SWILL ❧

Are you sick of the swill? Are you tired of low-minded meanness and boring mundaneity of everyday life? Then create your own reality and destiny: fast and open to millions of inner rooms like a castle—all with your name and the words "enter here, not a tear." As quick as a deer and so clear, just fast and there's only joy as you become King and Queen. No more mean for you've got the sheen--that means lean and clean: a genius gene's so alert and keen you won't need caffeine nor constant cuisine. Fuel with fruit and fat thy machine--you'll be as tough as marine. If the fruit, fat and fast is your daily routine you'll be serene. The next day you'll wake-up so happy--then take tangerine, some cheese and occasional green: that's our inspiring scene.

Are you sick of arrogant, derelict abusers? Become excellent your Excellency through fasting. Wake-up your whole town or clan today--show them the spirit of God by staying high in the head not

JUST SKIP DINNER

down in the gut a dead mutt with a big butt. To all your trouble's you'll say "so what?" as from your golden hut you'll proudly strut when free of that rut. Once a decade I feel lonely but the rest of the time I'm in bliss. Enjoy the wind, stars and sun then fast and pray your intruders stay away— you're no more their prey if you can enjoy this fastglorious bouquet.

◈PARTIAL FASTS ARE MIRACULOUS TOO ◈

The Bible shows many times how partial fasts work miracles to let all the food-prisoners go free with the power to BE—a discoverer, inventor, writer or whatever they're *meant* to be. With God's help it's all possible so make fasting your daily routine and enjoy. It's no decoy (like raw veganism) but like a new toy or magic wand it won't annoy but bring such great joy--the bad to destroy, the good to employ. For genius is held down but fasting overcomes enemies and all other obstructions—so you must fast to get the success power of Troy.

Rise up from the primordial depths. The feelings of rising in the morning after doing well the day before cannot be described--they are ineffable because they are of God. Suddenly you live in *elemental reality*: the elements of new moon and star, sun rising in the East, gentle breeze, balmy afternoons-- such are the calls of the wild. You're now a child: saintly and mild, no more so riled but truly self-styled--the audience beguiled as you smile. Let the wild in you rise up to tell the rest of the primordial depths into which you crept as you slept. Though in your past you wept, now you're the adept.

◈ FLY OR DIE–ENERGY MUST STAY HIGH ◈

To become the environment we must fast. To merge with the desert mountains, sun, and stars I must be in the head not the gut. A fig or cheese then become the universe again. Do you want to fly or die? Then don't buy the three-meal a day or low-fat lie. Fast

20

JUST SKIP DINNER

after fat, delete all starch and it's bye to your cries as you're never again so wrinkled and dry. They'll give you the eye with your thoughts in the sky with the angels—now that's high. If no more rye you'll be so spry with a beautiful thigh so give it a try then release a happy sigh.

To stay high in the head you must have nothing on your stomach and if you eat you must wait for digestion to complete then rise back up again, for it's a matter of where the energy is-- and digestion naturally drags it down. Due to *encephalization*—brain enlargement from high quality food like fruit or fat and fasting--the faster has royal privileges. *Now* you can be the longhead royalty you've only read stories about. There'll be no more drought, doubt nor pout so choose this new route and you'll let out a shout for whether you had a drinking bout or gout now you'll have clout once the bad is out. Now have some trout and then fast--now you'll sprout.

Bite the bullet, begin to fast and become brash—that is, *bold*. You now have the infinite power at your disposal for you denied the flesh in pursuit of the spirit and you enemies didn't. Let this be your fasting club and return whenever you need a "fast boost." Your spirit is loosed and now you look spruced, your dimensions reduced, your work now produced--audiences seduced.

⤚ FAST BOOST ⤛

That boost comes from truth—that fasting is the timeless method used by saints and genius throughout history: to resolve, find, reveal, solve, unite, transcend or defray, so do it today. Since fasting levels all playing fields it yields the pulling power of the most potent people in the race. This becomes your new program as the fast opens doors to these higher forces. You'll find yourself swimming in new realms filled with creative ideas "not your own" yet you receive the credit as God works *through* you. So from now on watch what you chew and you won't be blue. The others won't have a clue why your skin's like dew while they're filled with glue (out of every pore, goo). So get a new view and body too, bid bread and rice adieu so no fat will accrue and start anew—each time you fast it's your debut. These words I construe are intended to renew you. As you subdue your tastes untrue now you can *pursue*--so just fast, prepare for total success and wait on the Lord (for only He is true).

✍ FAST AND BECOME FAMOUS ✍

Fast to become famous in your town today. We're all meant to make a serious new dent so once you food-repent no more torment from cement—that was lament. Are you not spent? In any event let's now circumvent what we resent: what we *used* to represent. Fast and not just for lent, got my intent? What I meant was this: you know your bent? It brings a big present, much more than a cent. That is genius—God coming through the clear vessel--so in the clouds you will nestle with man no more to wrestle. Become divine royalty to those who know you—fast. Just start and your situation alights for by your intention alone it's better than Vegas as the faster gets "royal" privileges like favor in public. His face on a coin or the moon—in purity, he's the eagle enthroned in nature, the "well-known unknown". Fast and if you eat, just restart the fast. Never say "I've broken it so now I'll eat cake" but just start again—the *fasting consciousness* is all that matters my friend.

✍ REST BEFORE REIGN ✍

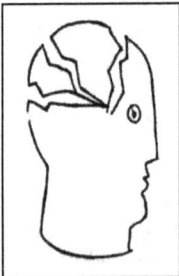

We need rest: one whole day a week and part of each day. We have infinite untapped creative reservoirs which won't open without rest. *Rest before reign.* Rest, then take the throne. Tired to the bone? Trying to be a clone when you could have flown? It's brought a big groan for it's all you've known--it made you moan when you could have grown, you could have *shone*. You lost your tone tied to hearts of stone so don't postpone the unknown: the past overthrown, you'll be happy alone with no need to phone. After fasting you'll smell like cologne, a busy cyclone.

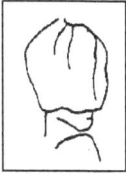

Fasting brings an incredible sense of well-being as something you've *achieved* and this relief is the necessary rest recharging the system. Indians don't work, they fast--for that attracts solutions and remunerations not coming from meaningless assiduity. There is good work—creative action inspired by God--and bad work—useless diligence forced by a darkened mind. So let me remind you to unwind, then no more undermined. Work and leisure intertwined is the best life to find for had you left food behind you'd have always shined rather than being blind. When you wined, mankind maligned you but now you're refined so no more confined.

✺ ANTI-GENIUS FORCES ✺

This generation can be mean-spirited due to toxicity of modern foods. The champ must be impervious and guard what enters his blood, mind and environment. Fasting is the way to go deep and remain unscarred, for wasn't it hard—getting charred while you sparred? You're so avant-garde yet they had no regard--they just sought to retard then discard. You were caught off guard but through the fast you're a die-hard so go one more yard—it'll fill your dance card as you say "I starred".

What makes certain people reach for things like this? It's for the few who can see reality *clearly*. What brings people down into derangement or degeneracy? Food, habits and other people. Genius must fast to detach disaster and no longer be cheap because we're so unique and deep they think: "creep." But we aren't those blind sheep asleep as over people they weep. The championship path is steep so you must *clean-sweep,* and *then* you will reap. Once clear make not a peep--then its your soul which will keep having made that giant leap. You'll need no more sleep and P.S.: you're no more the black sheep called "trash heap".

JUST SKIP DINNER

Have you always felt destined for something great? Fasting is the way to finally bring it out. Fast, get taut—show your face to the masses: it's now a lithograph carved on a coin and written on the moon, much better than looking a prune so out of tune. Don't shovel that spoon, can't you fast after noon? You say it's inopportune? Well just give up the saloon or resembling a balloon--be a Daniel Boone, not a goon.

❧CAN YOU REIGN TODAY?❧

Stop-eating. Stuffing down feelings *never* works but fasting to dissolve fear *always* works so the notion of fasting miracles will get a fat nation to stop-eating. I know the problems of living this life: worrying about bills and relationships, dealing with hurt and fearing aging but with fasting this is all made right by illuminating solutions--it's the way to God's power by dissolving the ego and all the world's falsely foolish notions. The fast is like lotions or love potions: it soothes the emotions which in genius are as deep as the oceans so don't go through the food-motions—the meal brings demotions but the fast means promotions.

Fasting gets past the ego-facade right to the Master of the entire universe. He holds your destiny in His hand and fasting gets His attention fast. It's a blast: no more out-classed, all our troubles have passed. Denial lifted we may be aghast at our past but let that all fade--look to the future so vast. Thank God the war is passed (all that flesh you amassed then being harassed with their lambaste) but now what a contrast: the enemy we've outlast as now we're *unsurpassed*.

❧LOCALIZE EATING❧

Fast each day by eating once or twice only: localize all eating into one part of the day (mine is before noon) then blissfully sail off into a no-hunger fast for the rest of the day (18-20 hours). Or you can continuously fast only *punctuated* by mouse-meals.

JUST SKIP DINNER

Either way you'll see problems dissolving, relationships healing, money matters soaring and no more life so boring as you see the blessings pouring. Now we'll be scoring—no more fat-storing, noisy snoring or tribal warring for as the fast is restoring the fans are adoring. I'm underscoring and imploring: fast to prepare for outpouring then start exploring.

Why isn't this a world rage? Because it's a Dionysian generation who wants to eat along with other short-term pleasures and never "fast"—a word it hates. It's also a Godless generation seeking peace in Self and psych-remedies for everything but God so this info gives you the edge as you begin a new life of *divine* (as the grapevine) fasting miracles. Each day, every hour no more to cower or need to devour for you're done with the flour--it made you fat (and those cravings made your mood so sour). Now my little flower be ready for power: get high as a tower and the blessings will shower--are you ready to empower?

❦ JUST FOR TODAY ❦

The question is do you have enough money, love, shelter or comfort for just this day? If so, don't worry and just fast this day after which the whole problem is over anyway. Get your mind off the problem and onto the delicious feast which is the fast. A whole new vista is about to open just by thinking of and *intending* to do it. Will your problem be resolved the way you planned or wanted? Maybe not—but you'll see a better plan (it is God's) or quickly transcend the problem altogether, now light as a feather free of skin like leather. The answer lies *within* and fasting reveals it or allows you to live with, like, or even *love* the problem. So the point is: start the fast (just skip dinner) and joy follows soon as you'll be so perfectly attuned they won't impugn. It's like a cartoon as you become immune to the baboon-buffoon for the daily world faster is a holy commune too high for anyone to lampoon. For you are a could-care-less happy holy faster after noon living in a romantic dream as you enjoy the castle of a sand dune or sitting on the moon.

JUST SKIP DINNER

❧ FAST TO FIGHT JEALOUSY OF THE GREAT ❧

Never tell anyone your method of fasting for it's none of their business and they'll all deny it anyway. Your *intention* is all that matters to the Lord so if you eat just restart the fast--it's the spirit of fasting that works here and it's just between you and God. As you will see it will change your reality to the highest you can be--so the fast is the key. As deep as the sea will be this journey with God and thee so take it--you'll agree it's the only way to be free: living without debris. Hear my decree: food is so bourgeoisie so be a gut-amputee then a busy bee--I guarantee a true jubilee. To the champion devotee: fast and be a retiree into life's potpourri, to the *highest* degree.

Do you have a problem on Friday? Fast for the weekend's sacred Sabbath fat-fast and you'll be amazed at a weekend like you've never experienced as in the short time of rumination you receive illumination. That's your fasting fate for it's with *higher* thoughts you have a date. Then you'll be great as fear leaves and no more hate, a clean slate. Your gene-trait can't wait—you must get it straight so drop the weight then no pain to abate. Now you can create your grand estate and there'll be no debate--for you're now first-rate.

❧ FASTING MIRACLES DIARY ❧

You're about to enter a cornucopia which is magnified through a Fasting Miracles Diary. Writing down every miracle as it unfolds and always re-reading when down will build faith in the Fast as the only device to receive His almighty power so blessings will shower. The fast opens the door to a new perception of reality, a clarity you've only dreamed of in fairy tales. Now you become a "discoverer" as inner and outer worlds illuminate as a kaleidoscope--this is our way to cope. The eater stays dense like a dope but you can have hope at the end of your rope for as we fast it's all down-slope, cleaning like soap. Fasting makes mountains into molehills so just start and all irritations drop away. For you're not escaping life like through drugs, but going *deeper* by taking on a discipline which is time-honored, ancient, perfect, God-made, spiritual, and *always works*. What leaves are the quirks leaving only our works which were perfect (though we may've been jerks). You don't have to do it for ten-30 days, just have a nice lunch then just skip dinner—now your guardian angel lurks.

✄ STAY HIGH IN THE LONGHEAD ✄

The fast brings encephalization or brain enlargement—the "longhead" marking the monarch, a sign of royalty from the beginning of time. It's the eagle, the symbol of freedom and exploration, the right of the regal so make the fast your profession, life, and joy--it's legal. The world is dense about fasting so here we're in bliss spreading the word to the mass to be led and hitting a hard wall instead. For that we were never bred so just fast, wait for your time to come and rest in bed. Avoid the bread (manna of the walking dead) for that's why you dread—you were wrongly fed. From monsters you fled but it was all in your head for your blood wasn't red as the white stuff spread making you weak as a thread (so upon you they tread). Well now you're in good stead by looking ahead, a *red* thoroughbred. Because it's not yet widespread you've a right to a swelled head.

✄ SACRED FASTING WEEK-ENDS ✄

No matter what you do fasting covers it with a gold seal of approval, endorsement and nobility. You may have feasted or partied and feel panic at your bad behavior. Fast to transcend all problems and become better than you were before it all began. Join the clan—you'll be a fan for it's easy to plan and an easy tan. Are you a man or a van? Fasting just this one day is worth a thousand years as you change with God's speed and step into the fasting reality where self-transcendence begins. Once there (beyond petty self) you'll never want to leave home again. The best example of forgiving transcendence is the sacred Sabbath fat-fast—the worthy week-ender separate and sanctified from the rest of the week. It's so un-bleak and very chic. Come join our clique then you'll look Greek at your peak yet still so meek. That's not weak so seek the sleek for you're no antique—you've got technique and a good physique and both are so unique!

Keep the week-ends special: begin the Fast which is the Feast on Wednesdays or Thursdays starting the most enjoyable Sabbath Fat-Fast after a completely appetite-suppressing (that means high-fat) meal. The two-speed week is three days business, four days God as the spiritual is magnified through a fasting miracles diary. Just mentally "take note" each time one happens or catch each miracle on paper to always re-read when down. Soon you'll wear a crown and as energy/talents are released in applause you'll drown as the talk of the town. You're now brown and renown--free of that frown looking so old and run-down.

✑ MAN A FAT-FRUGIVORE ✑

If you eat high-fat like cheese you'll be well-nourished and not crave culture food but if deprived of this wonderful fat you'll crave the infinite varieties of junk food and never be satisfied. Stay sane---take fat and some fruit to boot. You'll be so cute like a flute with plenty of loot to dine to the lute so take this route to get to the root, that's our pursuit. When finally free, just your own life computes so no more dispute.

The fat evens the emotions and balances the sweet while completely suppressing appetite, a necessary thing when eating so little and fasting most of the day. This balance makes the champ loving but firmly in control: he adapts to no one and is like warm steel as his environment reflects his personality which is then enhanced. Perfectly balanced and never craving he becomes alert, strong and beautiful to look *up* to. The eighteen-hour fast is always rewarded as your image to self, God and others is increased (like it never ceased). That crud was a beast filled with yeast and it made you feast.

JUST SKIP DINNER

Now that fruit, fat or fat keeps you greased you'll not be deceased because you decreased--*now* energy is released.

᪣ BECOME A REDSKIN AND THIN ᪣

In this first summer of the new regime your skin won't show a wrinkle and become quite dark from rich red blood showing through spiritualized tissues. That's everything you want, so go forth. I came into Paleo-Fasting when down in the dumps in the deepest darkest depression. I left the dungeon by eating some fat once a day in the most exquisitely beautiful surroundings. My life exploded and I saw the point of dietary fat: I'm never hungry so why eat? This is being abundant and rich so no need to diet-cheat.

I believe in reversals between fruit and fat as we can live so happily on just a little of the "HQ": higher quality potent foods like fruit and fat punctuating every fasting day. Due to protein-deprivation your immune system has been suppressed creating intolerance to many foods but now you're health will jump back suddenly so don't dismay, perfection's coming your way. Just find the fat that works and stick to that--as protein enhances immunity you'll be able to enjoy all previously-intolerable foods to your great delight. It's so easy as the fast deletes the fight and gives back your success rights.

P.S. An ex-fruitarian expresses his failure to thrive:

"Ehret the espouser of fruitarianism was not God. I too idealized the man until protein/fat deficiency suppressed immunity and my spirits sank. In desperation I took protein/fat and instantly came back."

29

2
MIRACLES FAST

*Daily Fasting will change your life forever. Longterm fruitarians
have health deficiencies but fruit and fat offsets these problems
bringing the bounty of fruitarian consciousness and the bliss of non-
deficiency.*

❧ HOW TO STAY YOUNG : ENANTIODROMIA ❧

The fast is stepping into another reality. It means
changing for good for you'll never be the same,
having separating once and for all from the
maddening crowds and the morbid past. Yes,
there is an escape--*within*--and it's just between you
and God who designed this final solution for man.
As you eat fruit, fat and then fast for the big change just say:
"I'm going to be entirely new in just three days, regaining the
appearance of youth." Soon you'll be brilliant and alert like a
sleuth.

Humiliating failure precedes greatest
success   as the fast transmutes the
burdensome fight to a "famous fruit".
This is *enantiodromia*--the inversion of all
systems as you become the head, no more
the tail. No more fail nor need to wail for
this divine device you'll now avail. You'll
be thin but strong like a nail not fatigued
and slow like a snail (nor for some as big
as a whale). But on a new adventure you'll
set sail so put out your fat clothes for sale.

30

JUST SKIP DINNER

Early readers who heeded my tale (sent through daily email) claimed they won the Holy Grail—it was behind a universal veil down an old Indian trail.

∽ OLIGIPHAGOUS FASTARIANISM ∽

 Now it starts--THE FAST. The one thing you've been looking for: the key-lock to success, the transition reversing destiny forever. The Fast is the step-function catapulting your life from the lowest to the highest. Start the day with juice and then soon have your fat—cheese, fauna or just peanut-butter: you'll be amazed how it suffices. Jumping into sweet fruitarianism is a big mistake so go easy as the fauna-fat brings balance and saves the teeth. We need *saturated animal fat* for B12, energy, fat-burning and glow—soon you'll be ready to show. A diet restricted to fruit, nut fats and animal foods is very high tech—no starch, that's the problem in fact. Simplify your diet by being *oligiphagous*: optimally existing on *fewest* varieties so the body can *maximally adapt at the highest speed.* I use grape and apple juice, peanut-butter, cheese and fish: such simplicity and ease is a refreshing breeze, for the less variety the less craving (and these are the keys). For thirst take your morning coffee and afternoon lemonade and eat the fruits you please.

 Take fruit to clean/energize and fat to stay in fat-burning mode with full appetite-suppression. I found I did best on juice when first hungry then just a little peanut-butter, then melted sharp cheddar or swiss cheese (avoid cheap cheese!) for the main meal, or fish salad with onion and olives. My readers also enjoy doughless pizza (see recipes on website).   Many love the tomavo, guacamole, red salad or doughless pizza with diced tomatoes and zuchini. With fat fasting you'll shine "high" during business hours and be rid of hunger for good. That's the main attribute of fat: full appetite suppression, and you can eat all you want of it—it all burns off if no starch in the diet (without starch you can even fry it).   Some will do best with fruit and fat, others do best with no fruit just fauna (like a cheese omelet every two days), some will prefer   just non-sweet fruit with avocado (fruit-fat), and others do best on reversal dieting between fruit and fat-fasting vacations on weekends or at restaurants. For those who enjoy restaurants twice a day, do an omelet in one and

salad/veg/fauna (no spud) in the other. Whichever way you choose, now you're going to win not lose. For the vegan and fruitarian shows *failure to thrive* not this cosmic mind cruise. My health jumped right back after a haggard low fat stint which brought the blues (it was bad news).

❧ FASTING IS THE POINT ❧

The main point is fasting which brings VITALITY--energy coming from least obstruction (digestion and bulk) keeping the energy in the head not the gut. Due to brain-gut competition the less you eat the higher and more perfect you will be. Because it takes so little to do the job fat keeps you steady and satisfied not dragged down by your gut. You'll be able to live on a little cheese or fruit in season, enjoying life to the max as you explore your territory and putter through your day. When you get sick of one just reverse into the other. The reversal diet between fruit, fat and fasting will enliven and enrich your life (it's your new paleo-nature mother).

❧ MINI-FASTING IS DIVINE ❧

Some of you will have a big breakfast and fast to the next day or over the week-end (mini-fast). Like my reader Pietro who after one omelet could easily fast for 60 hours and said "where will I go from here? This is so fantastic I can't believe it and refuse to discuss it--though everyone sees the big change in me." For the high-energy person just fasting until lunch is like one day and one day is worth a thousand years to God whom you'll begin to know. Soon you'll be ready to show.

The fast is your time in relationship to the divine. Try it and see as a whole new, higher world opens up from morning to noon with the highest. By the morning you've already been fasting twelve hours so take advantage of this edge. Just fasting each morning can do you worlds of good as just one new routine can change your life as these early hours are beauty and productivity combined. The fasting mornings are a kaleidoscope as everything seems "heavy" and "fat" with new meanings each minute. This is fullness at it's most exciting as hunches hit you like lightening and each day is more enlightening.

❧ SACRED FASTING WEEK-ENDS ❧

This is how we stay young--ageless--and totally repaired and rejuvenated no matter what our age: Fast as long as you can between meals, taking nothing in but distilled lemonade. This along with fauna fat will release an astonishing amount of human growth hormone which repairs, youthifies and makes the vessel perfect. Especially after age 40 when HGH is greatly diminished the fast completely offsets this natural shut-down, and beauty and youth is restored. If one is not interested in long fasts he should at least eat small meals and fast as long as he can between meals—for HGH (for a total selfix) is released with calorie-restriction as well as fasting, especially at night (just skip dinner to be tomorrow's winner).

Why isn't this a world fad? Because people are so trapped in lower food lusts they'd rather take pills and potions to feel they're accomplishing something without having to sacrifice the food-fix. Or they're so hypnotized by raw vegan philosophy and fear of the animal-on-the-plate they've become dense (the brain has atrophied). Try it now: eat fruit or fat then start a short fast. Now you'll go beyond your problems, remembering that pig-rat no more for now you've found a Prince. In a short time you'll go from a shack to a castle. First in your mind then in outer reality you'll be convinced (to all readers good health did evince and they've been loyal ever since). The more you get used to fasting the sooner hunger leaves each time you embark on the trip. You get so good that days pass by blissfully as you love the effects as well as rewards: a creative and strong ship. As these days are ingrained in mind you so value the fast as a happy-healthy-Hopi vacation that eating may even bring depression.

❧ RAP-ORATION ON TRANSFORMATION ❧

JUST SKIP DINNER

I've made it my vocation for it's a matter of vibration, nd oh--the sensation. I'm galvanizing a whole nation to guard that ration, for eating like that is not your station (all from desperation) and the resulting constipation is a failed destination. So let me impart information for your education then just fast for the full integration. Without hesitation embark on this operation for it's your medication, a skillful navigation and mind-transportation (that's your meditation). Like a pure carnation you'll be the divine creation--all else damnation. In society you'll be causation with all strife's cessation so come join our daily celebration with full compensation. It'll be pure fascination so expect constant elation.

If you hate the uglies, avoiding temptation is your salvation. You must face the frustration to be a cute decoration--make it your occupation in sweet isolation. Without reservation I'll say it's your self-realization, true relaxation and the greatest preparation. It brings inspiration--are you ready for innovation? To be thin, rich and thriving is your motivation and there's no need for explanation for we're a new generation. So without trepidation drastically change your situation and you'll have TRANSFORMATION.

◄❧GOD'S FRUIT ❧►

Fat-Fruitarians do well on dried fruit (raisins and figs) and non-sweet fruit (tomatoes, bells, cucumbers, squash, avocados, lemons) but the real punch is from fauna: animal fat. Without that you'll go flat and even start combat. For it's an evener, a balance to emotions and soothe to the glands—this kind of comfort is what your brain and destiny demands. *Then* fasting makes the next mornings blissful. If sugar doesn't spike insulin enjoy juicy sweet fruit too but eat with caution: it makes many blue. Fasting is such a wonderful balance to God's delicious fruit and fat. Whatever the problem, fruit and fasting kicks it out and reboots the machine. It means lean (if without the bean) and it brings out your genius gene. You'll have clarity without caffeine as fat illuminates the scene--a necessity of the King and the Beauty Queen.

JUST SKIP DINNER

Reversal dieting between fat and fruit makes for a happy, satisfied and blissful life. Fruits cleans to the core bringing exhlaration and fascination with every moment. The look, the feel, the experience of the fruit-fast cannot be described. In a cleansed body, the sense of euphoria is ineffable as dynamite fruits like figs, raisins or grape make these times unforgettable and blissful. Then if followed by fat, glucagons is elevated and one enters the energetic, cornucopic and satisfied fast all afternoon and evening. This is truly living like a king. I have eaten fat with fruit for many years and remained very strong but my life is mostly about daily fasting which the fat makes easy and safe. As a lowfat failed fruitarian I was just a desicated fatigued waif.

❧ELIMINATION: BE NOT PASTY❧
Starch, or Energy off the Charts?

One of the wonderful things about the combo of fruit and fats like avocado, nuts and cheese is the regularity of evacuation--the cleansing action of the fruit combined with the roughage and colon-happy fat is unbelievable to those making the diet change. The journey is maximized by this daily cleansing-elimination so each day one has better perception--lucid sights and sounds. As a pure fruitarian I never had this type of efficiency so to those thinking "cheese is binding" let me say that the grease even creates diarrhea. Good--let it occur! Watch luster return to your fur for the colon needs fat, sir. What constipates is cheese combined with starch (it's the latter that blocks, making one parched) vs. energy off the charts. The absolute ease of daily evacuation from higher paleo and the moist sheen to skin and hair indicates it's superiority as a diet (ask any vet). Low fat diets are a skin and hair-threat for both you and your pet. Regularity is so important on this journey and it's optimized through fat (esp. nut), fruit and the afternoon fast which initiates elimination mode. Fat for the brain, heart and glands is the form of energy the body prefers for it is not glucose that energizes, it is *ketones*—these are the cures.

Obstruction (sin) misleads to tangents and bad actions. Through the fast we contact the source which acts as a creative spring to the true self and the destiny we're *meant* to have. From that point on just by being yourself you stay ahead, cleansed of the past. There is no more need to cling to and rely on compensatory symbols of prestige and status which corrupt the creative spirit. Your talents start to blossom just as you're relieved of the dry, caked constipation which comes from the lack of fauna fat, dry foods, food inventions and never fasting.

❧ NO PAST : ETERNITY IS THE FAT MOMENT ❧

The fast obliterates all bad memories of the past or reticent apprehensions about the future. It smashes through self-doubt and says "I *can*. It *can* be done." It illuminates our destiny which can be realized just through this new accomplishment alone. The wonderful new sense of purpose overrides the feelings of powerlessness, hopelessness, has-been-that-never-was status. Whew! What a relief to move on from the corny past to the miraculous future: being on *top* for once.

When you fast time collapses. Simultaneously you sit in the past, present and future--the "fat" moment. Once you enter this state you will never be bored (or lonely) again. The more quiet and non-chaotic your environment the more pithy and kaleidoscopic your fasting experience from deep within. This is: ETERNITY. The more clean and clear the moment the more infinite the consciousness, so clear your possessions out and delete all distractions like TV and talkative hangers-on. Now, you're ready to embark on the inner journey.

❧ RECOMMENDED FOR REVERSAL DIET ❧

Anyone beginning this diet should read my book *Daily Fastarian* and *Champion Guides*. For the fruit cleansing background read

JUST SKIP DINNER

Ehret's *Mucusless Diet Healing System* and then *Rational Fasting.* It is dangerous to go right into fruit-only for long periods especially without a long transition. Then I recommend you read Atkins *New Diet Revolution* and Eades and Eades *Protein Power.* With reversal dieting between the two, you're getting the best of both diets while offsetting the shortcomings of each. If you don't want to read these books you'll get all the info you need from mine: eat fruit, fat and water with lime.

◆ THE SAGE SAYS ◆
"It Takes Fat to Fight Fat"

This culture is fat-phobic yet more obese each day. Dietary fat is essential for low-fat diets create depression, suppressed immunity, ugly skin and dull hair, accelerated aging, bloat and obesity. Because fat foods are substituted with starch and sugar insulin is elevated which lays on fat. Dietary fat in contrast elevates glucagons which puts the body in fat-burning mode, elevates energy and suppresses appetite. It literally takes fat to fight fat.

37

3
MAGIC DOOR

The Fast is the Magic Door of the Superior
Man—So They'll Do Things his way.

Fasting is the magic door into a new kingdom with you the King and God as your Director. It is impenetrable, irrevocable, irrefutable and very impressive. It is yours while fasting for you're above all logic laws and the losers around you. Do what you want and trust your instincts for they're always right as long as you fast, then hold your head up high and tell the simple truth—you must or combust. Thrust your chin out, be yourself then in God you *must* trust.

There's always a last straw that makes you recognize it all and never go back. If this is your first mini-fasting day you are double lucky for even serious diseases can be cured by this simple method of just eating lunch and/or breakfast. Eat only once to be happy all the time and achieve your goals, for suddenly you've released the system and your work will come out. Nothing can hurt you now for you've found a brand new life and well-being through the magic door--for fasting is a mysterious adventure you step *into*. Like an old romantic movie gone is the blue, so fast after frugal meals (like cordon bleu) for only this is true. Are you ready for destiny and bliss—the champion fast crew? You'll be saying "I'm so happy with my fruit-fat-fasting stew—only FFF is true."

JUST SKIP DINNER

❧CHAOS OF THE CLUTTERED CARNAL MIND ❧

The eater whose brain is in his gut is entombed in millions of superfluous and trivial details—it's the chaos of the carnal mind. He's so caught up in minutiae that decades go by without ever seeing the "whole". Not so the faster: living in the head he has the necessary clarity making him a natural leader of the trivial mind. The sheep need him to direct the show lest they go so low. To their ideas he must say "whoa" as it's *his* gifts to bestow to all those below just like centuries ago. Their line he can't tow (so much mundane in a row). If they're in power he must eat crow—that's too severe a blow so as meek as a doe he becomes like Thoreau. He must lie low for their status quo is not apropos (for him the pro it's like skid row). But on his plateau his talents will grow then he comes back with new powers: "hello". Fasting and solitude makes him all aglow a leader so beau—he's back in the dough. So no more to and fro it's on to power we go. While the flock eats bread dough with money to owe he's where he belongs in his chateau. As a universal archetype—not a mere cultural stereotype—the True Leader sees the predictable outcome of things-through-time (the "long haul") and thus he prevents disaster. That's the faster—he's the Master like a true pastor.

❧ THE FASTING LEADER ❧

Fasting makes one humble. You need this humility to be open to and access God's infinite power as your puny ego-- the lower mind--blocks. Never trust the natural mind for many of us are borderline. Have your enemies laid a trap for you? Fast and watch *them* fall into it for it results in great boldness from a higher power (necessary to make a moral dent). That's the faster's bent for he's got the God-scent. They give more than a cent for his consent for he's been divinely sent. What he says makes them content—it's like a tent protecting from torment so let him vent: When he says "repent"

39

see it as a great event for he has good intent. For when clear of foibles and fat we start to invent for a great percent. So don't resent the fastarian leader for he'll lower the rent and all troubles he'll circumvent. Though he's solid as cement some may mistake his boldness for arrogance when actually it comes from greatest humility. Simplicity of speech (necessary boldness) is a sign of guts, grit and self-assurance so to his edicts have adherence. Are you this leader today? Fast, and they'll want to do things *your* way.

❧AVOID INESSENTIALITY ❧
Clutter, Debris and Superfluity

The faster freeing himself from the frivolous and superfluous must have a new slogan: *avoid inessentiality*. He stays in the pure "quintessential", the prime essence of the thing or goal, so him you should extol. He sees the whole: that's all of life in a bowl (so good for the soul). Eating takes a severe toll as the gut forms a roll and one gets ugly like a mole. But with fasting you're as cute as a foal with eyes like coal and so happy like being free on parole. Just taking a stroll is so much fun with such control as all your fears the fast consoles.

It is the greatest rest to gain freedom from the trivial. It's entering bliss and then the longitudinal vision defining the True Leader. And so we come to the slogan *rest before rule*. It'll be so cool—like dipping into a pool--a true vocational school as the fast is a shaping tool erasing the fool. In rest one gains power by contacting the elemental reality made by God: In the fast the elements like sun and air thrill the tissues--one can hardly describe these physical-etheric delights! A million inner castles illuminate along with the electrical vibration, the elongation, the sensual excitement and consciousness of eternity through the sun, moon, stars, solar radiation and nitrogen. It's our secret den: I can't remember when I was so handy with the pen. It's higher than Zen, times ten.

❧ AVOID CONFLICT ❧

JUST SKIP DINNER

The Muslim or Yogi says "don't bother me—I am fasting." He learns early that fasting is the time to abstain from all stressful thoughts and conflicts. Fasting must include not only food but bad associates and silly-stupid-superficial-strife-filled thoughts, sick systems and cycles of sin. To gain great power be nothing but a vessel for God's spirit to come through--for here is where your greatest abundance, notoriety, prosperity and continued creativity occurs. Forget yourself to get it all, but be bold or people will stay dense--in the dark. Be like an ark, the fast to embark. Be not a shark with a stark heart and loud bark but make your mark—in them light a spark. Conger up a beautiful park, by your mere remark.

✑ FROM CLUTTER TO CLARITY ✑
Fast on All Obstruction: People, Habits, Food and Junk

Fasting is the biggest eliminator. The longer you fast the greater the elimination when you eventually *do* eat. Fast and you'll actually elongate as your head reaches for elemental eternity--everything "moves down" just as your spirit rises *up*. You suddenly feel so much better as the sense of puffiness or constipation ceases. Fat's gone to pieces then suddenly decreases--feel the energy as it releases? Now watch as it constantly increases. We all know that puffy blocked feeling. Fast and let all energy pull upward--now you're part of the Tao so to destiny you can bow. Wipe the sweat off your brow and forgo that chow for it makes you as big as a cow but if food you disavow—wow, as sweet as a kitten: "meow". Now fast also on storage—the literal junk we keep in our lives. These pieces of possession clutter the mind and even paid storage obstructs clarity. The faster must become obsessed with elimination on all levels: go through each drawer, crevice and name in your phone book. To become King become simple: free of strife, the superficial and the sordid. Your every move is recorded so those sins you're boarded cannot be afforded but the fast is awarded—just watch as you're beautifully rewarded.

✑ EATING FOOLS VS. FASTING RULES ✑

41

JUST SKIP DINNER

Most Westerners are either gabbing or gorging-- they are "eating fools." Fruit or fat requires one meal to suffice bringing a balanced poise to the emotions while the fat suppresses appetite. When the fat-phobic insist on deleting fat they eat and eat but are never satisfied: when they're at a restaurant they talk of other restaurants. They can never comprehend the higher fasting life of the fat-fed brain of conversing with angels and the Great Mind—all the genius and sainthood that has ever lived. In the inner journey you must get past all this and fasting combined with fats is the way to do that. Jumping right into fruitarianism releases too many toxins bringing uncontrollable cravings and even anger so go slowly and enjoy your new ride: enjoy the manifold tastes of fruits, fats and then the fabulous fast. Into a wonderful new role you'll be cast.

◈ AVOID INESSENTIALITY: AIR SUPERIORITY ◈

You have heard it said "he who does not work should not eat." Well I say to you that "he who does not eat shall not have to work" for it is attracted in *naturally*. Having first eaten right the fast electrically conducts God's magnetism and you *snap* to your goals. Fasters are supported "on high" and because they're high in the head they are closest to heaven--smiling through any battle and winning hands down every time. Due to the fast they're ever in their prime: water and lime makes it so easy to climb. As one shifts from digestion of gross nutrition the body automatically re-adapts to getting it all from the "superior invisibles" like air and sun. They have "air superiority": No matter how hard the battle the faster confidently smiles into divine blessing. You don't need a gym to be slim and trim. Eyes no more dim nor life so grim but joy to the brim so go out on a limb for it's not just a whim as into the sea of destiny you swim.

◈ AVOID PALTRY PEOPLE PROBLEMS ◈
Logorrhea

JUST SKIP DINNER

When eliminating non-essentials be sure to include words and people: most talk too much when only ten percent of their words would suffice. Let words be a planned device, very concise (not like rolling dice) but precise, the listener to entice. That's my advice: loquacity is like ice: it just isn't nice as it irritates like mice. And you'll pay a price for that vice so let your descriptions of life be a *slice*—don't say things twice. Increase your verbal powers with the "essentiality slogan": *less is more.* This applies to words, clutter and close associates in whom you confide or ask for "advice." For you must get away from people's control for they take their toll--as you miss the moment. It's the Godless losers that use psych-labels to invalidate *you* so fast and feel no remorse about leaving these users behind. Fret not these evil-doers--they're only displaying who they are so you can be free of them. Just by getting involved with the wrong people you can find yourself in poverty and closed off from your destiny, work and talents. One must become *defensive* if he has lower elements around—a stance which blocks the creative for sure. Conversely the minute you cut them loose your destiny will blossom--it'll just be awesome.

✌ SPIRIT OF FAMILIARITY HOLDS YOU BACK ✌

People may respect one's leadership at first but soon a "spirit of familiarity" arrives and respect is lost. The good leader for the sake of the people must remain a bit separate: *aristocratic reserve.* A general rule is: *going solo perfects all work.* Then it's just your fulfillment of the Creator's divine plan *through* you. You won't be blue for it's your due. We're very few: the rest haven't a clue that it's all what you chew. Their ego can't accept that—they think they *are* what they *are* and they don't have to "work" or quit their quirk to be a star. Now your time's overdue, so think "all else untrue". Your youth will renew just like shampoo for the future's in view. Your funds will accrue—like never you knew--so take my cue and you'll be rid of the flu or your cells filled with glue. It's a catch-22 that pasta-in-the-stew. Don't you want the true? Those starches are taboo. or hunger will not subdue in a body askew (from side to backside you grew). As you become new with FFF you'll have a new red hue—so your problems are through.

43

JUST SKIP DINNER

You're a refugee from the human zoo but now destiny's ready for you to fall *into*.

❧ ELIMINATION INTO PERFECT SUCCESS ❧

The obstructions to genius are people, habits and food. If you're a food-addict, dissolving *this one* problem can suddenly dissolve all *three* obstructions. By eradicating the habit of eating, the food issue is also solved and the controllers (sick systems) are dispersed. Sin brings control because it creates the shame enabling it--they go together. To stay free and fine, avoid idiots! Coming out of a frightful persecutory past your slogan should be: *clean sweep precedes success,* for it's really been a mess. The excess, the stress—you must confess it ruined your dress since beauty comes from happiness. When you transgress, your talents suppress. You had no finesse so who'd want to caress? But I digress—just fast and take a recess: this stops the regress and all that distress for this genius you possess you can now express--it's not a guess, it's your *new address.* The throne you'll now possess as God will bless, so on people no more obsess. Instead, the whole *universe* you can access.

❧ LISTEN TO GUT BUT DON'T FEED IT ❧
Avoid People to Not Heed It

It's not only *what* we eat, believe me. Do you sense evil when with certain people, a feeling that registers in your solar plexus? Eaters dull that warning with food but fasters heed it and get away--fast. They know to stay entrenched means their awareness slips unconscious as they become dense and dull. For the only way of "fitting" is by *slipping* into blind addiction, even to the oppressors themselves. Some people are like pit-vipers: snakes that "dig up history" to keep you on the defensive. It's so expensive staying near to the offensive, for you're so comprehensive you must stay pensive. Just be apprehensive for their damage is extensive. Fasters must stay free to stay fine—to *lead* people, not be victimized *by* them. You're a gem so hem in that crowd from which you stem lest they continue to condemn.

44

JUST SKIP DINNER

You must keep the pagans at arm's length. These are those deniers and gainsayers of your very soul, spirit, speech and search. They tell you "what is what" like using psych-labels to invalidate and collude against you. Though your own perceptions seem strange hold on to the truth--hold your head up high and follow your own instincts. People always want to give you a plan or perception—*theirs*. But it's *your* map—adapt to none! Then you'll have fun as your work's not outdone. It's like removing a ton: you've won if old posts you shun. I'm not saying be like a nun or using a gun but rather knowing you're the one— the race you must run with nothing left undone. When the fast is done you've only just begun as you become the leader so take no advice from busybodies, boasters, the brazen, brash or banal.

ᘒ EXHAUSTION: A SIGN OF OPPRESSION ᘒ

Exhaustion is a sign of oppression--the result of letting inferiors in to rain on your parade through dense misjudgment from suspicion and envy. During the fast you realize: you did nothing wrong for they turn against the Great for doing *right* (you just went beyond their comfort zone). They just want a clone right down to the bone for over their heads you had flown. It's all they've known as they chat on the phone and it's made them like stone as all night they groan (as foods sins atone). But if the fast you don't postpone you will have shown. And you won't be alone—we've all grown so we too sit on the throne. We've got good tone! Now are you prone—do you want to *own*? It's like getting a free-loan as your future's full-blown.

If you "bought" their reaction, learn that people cannot see above their own level: the dense hook to the herd mentality: they fall to the social view which is cruel and false. Yes it takes genius to recognize genius. With time in fasting it dawns on you that misperceptions come from jealousy and disorder, not from God. Don't get drawn into the vortex— reject all these sick systems, sin cycles and silly slanderers. Fast and they'll dissolve as a vapor—as weak as paper, even the raper.

⇜ SELF-ESTEEM? ⇝

Much has been said of the need for self-esteem--that we can somehow build it just by *wanting* to. If you're eating like a hog or hanging out with hogs you cannot have the self-esteem of nobility. Nobles are above the swine consciousness of the herd—it's never preferred and so absurd. Fast to go beyond the problem which either dissolves or takes on a positive dimension. Suddenly it's of no consequence and you'll have the guts to lead. By refusing to frolic with fools you can lead by telling the truth that hurts. They've treated you like dirt--it wasn't a flirt their plan to convert. That sugar-desert is sure to pervert as it makes you inert (hardly alert). Those grains are so dry you must now revert and avoid all dessert. The low you must subvert to reach pay dirt: In your work you really did exert, so now all systems invert. It won't be overt for you'll be on top—*a very rich squirt.*

Once free of fakers your fasting life can take off for at the point of elimination of non-essentials the past divides from the future. The past was hell: treachery, gossip, back-stabbing, hatred, anger, maliciousness. The future is heaven: prosperity, popularity, love, creativity, acclaim, recognition and complete happiness. You'll reap as God bless. Yes the past was a mess: you were the less under such stress. But now you must confess: without the fast it was *you* who blocked your total success. Take hold on all levels for your ship is about to come in. With you God will dwell as he wipes away all your tears, shames and guilts. The former things will pass away as He makes all things new: You'll say "thank you" for it's like shampoo or a new debut. Our sins bid adieu, no longer askew and payouts accrue. Take your cue—join our fast crew.

⇜ ACCIDENTS/MISTAKES & OBSTRUCTION ⇝

Do you realize how many "accidents" occur from eating? It's all from lost conductivity to the source. To stay in sync—perfect flow with continuous miracles to your benefit—you must be in the head not the gut. Whenever you sense so-called "critical days" write a recipe: What did you eat, who were you around and what did you let in your precious mind? Be careful for there are serpents all around. They don't make a sound but keep you so bound. You're so profound they were confound so your problems did compound. But now the fast you've found so you'll shed a few pound. It's a giant rebound, being renowned. Your fans will surround like a theatre-in-the-round.

We all make mistakes. We all make asses of ourselves. We all have occasional lapses into insanity, faux pas, slips of the tongue, embarrassing social hot-spots. That is the nature of the human but the fast eats it up. Yes because we fast all the social sins and sly sadistic sillinesses are smudged out. The world is down in the gut marked by a mass of minute memories. We however are so far out of sight that no one remembers the saint as a sinner. As soon as the sun hits the skin of the faster there is total forgiveness. What petty problems we people portend: to our sorrows we bend with no money to lend (that was our tend). We thought life was at end with ne'er a friend nor dollars to spend--now don't pretend they didn't condescend or you didn't defend. Comprehend? You had to depend: like kissing-up needily to "befriend." You had to attend to them, a very bad blend. Don't you want to mend or start a new trend? Then just *transcend*: fast and you'll ascend, for when fasting (on people and habits, not just food) you'll fly above all these temporary trifles or silly insanities. You must lose respect for human nature to gain a new stature, my friend.

❧ NO RECALL: SAINT OR SINNER? ❧

The eaters age into weird caricatures. They are the fodder of cartoons as the body warps into age-specific deformities. How sad. It's all from food (the bad). The constant eating fills the poor intestines and reflects on the face. Not so the faster. He ages into glory and increased beauty and bounty. He has learned from his mistakes and this reads on his face: a beautiful carving, not a caricature. He's learned the cure: he's got the lure looking so pure. I tell you the fast is sure, like a constant world-tour all made for you--so endure! You'll no more be obscure but rather an entrepreneur for sure--finally secure. There's nothing more sure (you'll have *demure*).

The eating fool dies a beggar while the faster is the possessor of royal privileges with the power of Caesar in the body politic. Fasting is the way to savor favor in society, become the leader and a beautiful exemplar to look *up* to. Fasting accesses God so all needs may be met. Fasting is the way to get out of debt and no more sweat—so don't you fret! You've got power to get: you'll be well-met (it's a good bet flying with the jet-set: far better than the internet). And yet you feel a threat just because your stomach's upset? Be a fast-cadet and never forget how your silhouette will look (like a statuette). Even your fat-fed pet needs no vet (that's all wet). You regret? Do this to off-set: Eat fruit and fat at the lunch-set then fast for a day: now the stage is set, so bring out the string-quartet.

⮋ FAST WALK IN NATURE & FORGET TRAITORS ⮋

Walk, sun and fast to transcend all sick systems and sin cycles keeping you down. When outdoors live in elemental reality of sun, moon and stars: this is God's Eternity. Fasting is heaven *before* you die. Sure life has problems but fasting is perfect precision and protection, a problem-free existence. Any problems the faster sees as great challenges: an opportunity to strut his stuff and get his spiritual sporting blood up.   The faster sees all problems as just happy challenges to show off his skills. Take cheer—above the gut lies "full" protection. Transcend "filling up" to find true fullness instead. Tired? Just go to bed but eat no bread for it hits like lead. Much better a fruit and fat spread without a shred of starch--on such you'll not tread lest you want to be dead. From your foes you fled so no more to dread, for now you're well-fed and it really encephalized your head (now you'll seem well-read). Your blood will be iron-filled (so *red)*. Now abide what I've said: to the fruit, fat and the fast though be wed.

⮋ AUSTERITY HAS GREAT REWARDS ⮋

JUST SKIP DINNER

If you ever need a healing in your family or some other crisis, fasting and prayer can fix it if. Self-denial has tremendous power and austerity has great rewards. Then it's bye-bye to blues, ne'er a cry nor so dry. All people die but we *can* defy age so don't be shy. No potions to buy just think of summer in July. What I imply is that we must rely on the fast to get God's supply—it must *underlie* every day of your life. That's my rallying cry so give it a try. In the blink of an eye you're brighter than florescent dye. Now we'll see eye-to-eye.

Indian yogis laugh at Americans whose energies are always down in the gut and who are seen as rather stupid as a consequence. They talk and eat constantly—gluttony and loquacity tend to go together in what is called "Epicureanism". While the Easterner tends to be quiet, silent, spiritual, polite and dignified the American over-eater is a fool. Sure we need to eat to stay alive: one long luxurious meal a day is great, first-rate. But fasting brings nobility once having nourished the vessel rightly—lightly, that is *knightly*. That fat so unsightly will now be wrapped tightly if you just eat rightly. That's so politely.

✄ BRAIN-GUT COMPETITION ✄

The work on brain-gut competition is revolutionary: The less digestive work the more energy is released to the enlargement of the brain. And thank God for I always sensed all that three-meal eating and continuous fullness was not right. The huge defecations of everyday man is subhuman. Be a lady or a gentleman for whom only one high quality meal suffices completely. With fruit and fat it daily evacuates ever-so neatly and that is sweetly. Eat your one main meal then fast all day in silence, that is discreetly.

JUST SKIP DINNER

It has been noted that Ghandi's diet was fruit, nuts and milk (lacto). Lacto-Fruitarianism and the yogic fastic path is very prevalent in India and entirely spiritual: As the energy stays high in the head not down in the gut we rest in a much "higher" ascended reality where everything appears to fit together perfectly and life is continuously blissful and happy. That kind of dedication and self-denial has benefits way beyond what Americans can see. If you're in need for health, wealth or svelte just fast for it--you'll see opportunities abounding. It's a real grounding the results so astounding: your foes confounding, your new life compounding with opportunities surrounding. So now I'm expounding a new theory resounding.

4
THE DAILY FAST

Ramadan-Buddha-Sabbath. Let Daily Fasting Return
You toFfactory Settings—As You Were at Birth

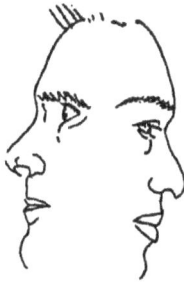

Fir all animals, daily fasting clears up major diseases. Daily fasting for the rest of your life will bring ecstatic joy to each day as you see your life continually changing for the better. Every day you'll see a younger face in the mirror, your elimination will work better and you'll crave the adventure of the coming new day--not food. The one thing I admire about Islam is the tendency to fast at all situations. There are basically three types of daily fasts: the *Ramadan Fast,* not eating during daylight hours; the *Buddha Fast,* not eating past noon and the *Sabbath Fat-Fast,* a 60-hour fast after one fat meal. Since according to Atkins 70% of Americans are insulin-resistant the importance of a "lowcarb vacation" must be stressed. Ehret himself stresses the danger of jumping into the all-fruit diet as it means tooth loss, craving, binging, moodiness, irritability, fatigue and crashing. Adding fat and occasional greens is the answer to disaster. Nature's changes are incremental, progressive and usually slow but the daily fast will speed it all up while you remain happy in the transition like a pup. The fat-fasting and fruit diet will make you feel so wonderful --it's hot stuff though not tough.

❧ MENTAL TRICKS: FAST FOR A VACATION ❧

51

JUST SKIP DINNER

When you fast, get excited. Feel as though you're "stepping into" a whole new reality or embarking on a wonderful vacation. It's your own fantasy you're creating here. Write down--and get specific--about everything you want from this fast: health, wealth or svelte. Keep a fasting miracles diary and write down each miracle as it occurs. You will be amazed at your resulting spiritual growth. Stomach pain? Bloat? These too shall pass as the area cleans out. You'll see how greatest pain precedes greatest beauty as all water and superfluity eliminates and the face reflects the gut. Take joy-- this is the most wonderful and productive time. Your life is about to change drastically and fantastically so endure it enthusiastically. Do it once and you're hooked for life.

✄ RAMADAN FASTING ✄
Bright Business Hours

Ramadan Fasting was my routine since sixteen. The idea of eating in the secret dark hours and then spending every minute of sunlight in the holy sacred fast—the business hours when I really *needed* that power—has always been an incredible device. When I do this I will begin the day with a *little* melted cheese with tuna, then some peanut-butter or a few nuts at night. Everyday is a miracle day with me and I am strong, happy, thin and creative. I am never hungry for I begin the day with appetite-suppression which fat provides--it is so comforting and satisfying: The morning meal— the Suhoor—is like a rocket firing up and then the afternoon's fast- -eat the sun (gotta run.). The more one fasts the Ramadan way the more he learns to expect daily miracles and then the more anticipatory these predawn hours become. One should eat in prayerful anticipation and glee for this day will soon benefit me (and everyone I see). Fast and just do your work as the light shines on and brings it all to happy completion. With energy depletion HGH is released—for fat and age-deletion it all goes out through excretion. Follow this wonderful fruit and fat plan: you won't ever be hungry and with the fat and fruit you'll soon look Grecian, as brown as a Haitian.

Whether for you or your dog, the appetite-suppression of fat works the same way. Some Muslims complain of having to endure this fast for religious reasons: It's unpleasant because they don't *eat right first* to make the fast palatable. The fat in the morning fires up the engine into full force and energy while suppressing the appetite for the whole day. Many of you will do best with a cheese omelet in the morning, but for those wishing to stick to fruit for the transition eat *tomavo* or *citavo* (see recipes online) in the morning and then in the evening fruit or a few nuts to eliminate all residue. It's all what you chew as to what goes through. Its just two: fruit and fat is our new stew and it's delicious too. On the fast you won't be blue as you were when you stooped to starchy food--as so wide you grew (what it put you through!) But now it's a *new you* so you'll say "Lord, thank you."

⤴ FAST STEP INTO BLISS—IT'S GOD'S KISS ⤴

You've got a new field of view and a déjà vu. You'll have a spring in your step like a kangaroo if it's just fruit and fat (like cheese fondue). All else but fruit and fat are so untrue for that's the highest quality (HQ) paleo stew--especially with the fast--so *pursue*. None will outdo for it's something you step *into*: a new worldview. Take my cue it's better than all that you knew as the food-fix farce really changed you: It's OK to tear into a cordon bleu but starchy grains laid waste to you. As you review and carry it through you'll see there is none equal to the fast after fruit and fat to be as strong as Kung Fu. Test out your case and find what fast and food suits you. I found all but fruit or fat leaves ugly residue, but you must find the unique proportions of fruit and fat—*only that is true.*

JUST SKIP DINNER

As you fast you'll see the more clean the intestines the less hunger experienced in the morning. It's a matter of mourning your past eating habits and heeding my warning. I made such an interesting discovery after eating some fat when first hungry in the morning: I was never--ever--hungry. Two decades of hunger were gone! The cravings for salty greasy fried chicken, pizza or chille rellenos were finally gone—fat-needs met, I could separate from the throng. I even gave up most fruit, with just a little melted cheese, a nut and some apple or grape juice. I had finally made the transition and saw Ehret was right when he said "hunger comes from wrong food" making us blue. Adapt to eating as *little* as you can—you'll be astounded to see those hunger pains aren't true—it's just the food demon luring you. Note: the demon of bulimia is the most ugly green thing, so *please* avoid that too.

✺ OBSTRUCTION BE GONE ✺

So you wake-up angry? It is likely this came from someone you were around, something you ate or thoughts you let into your mind yesterday. Do you wake-up groggy? It is likely you ate wrong the night before so try skipping dinner and you'll wake up happy and lively. Have your fatty breakfast then look forward to a day with God. Anger leaves fast in this awe-inspiring day. Just by intending to fast you'll see a balancing to all emotions and a release of wrath and irritability. Usually that anger comes from insulin-elevation—sugar and starch—which causes constriction and bloat (you feel as big as a boat then hide it with coat and you sure can't float). Take note: Give up the oat. The paleo-fast will promote while starch or sugar will demote.

✺ BUDDHA FASTING ✺

Most people choose the Buddha Fast: not eating past noon. You can just eat lunch or eat breakfast and lunch about four to six

JUST SKIP DINNER

hours apart leaving 18-20 hours of fasting for the rest of the day. I recommend fruit juice or tab. peanut-butter in the morning (when first hungry only) then after fasting as long as you can enjoy your reward: the most delicious red salad, guacamole or tomato melt (doughless pizza), omelet or fish—this fauna is your sword! You may enjoy a leisurely lunch or just have four hours of "food time"—not your meal but a grazing period enjoying these paleo foods while pondering, puttering and picturing your beautiful and bountiful future. What a wonderful Italian plan--localizing eating to a short span then leave the day for the clan and contemplation, exploration, meditation and all other creative benefits of fasting (even getting a tan). So much is accomplished in just twenty hours and done daily the results are drastic and cumulative. You'll "clean to the core" a little more while your person they'll adore. The morn and afternoons will never bore, it's like opening a door with miracles galore. It's a way to rapport, all functions to restore as it cleans to the core. All day you'll soar---it's a real *esprit de corps* so come to the fore. You've been a prisoner of war: the flesh became your carnal mentor making you *just want more*. That's not what you stand for and your gut got sore as it fat-stored. It's like a trap door ever-ready to roar and man it got hardcore. So you sought diets—a chore you began to abhor, so I'll say it once more: this isn't you anymore. I'm giving you the true score for a new décor. Just start your fast and be ready to explore.

Each day you're approaching the True Self and God who designed it. All these fasting benefits yet you're never hungry! To have a delicious lunch by noon one isn't hungry all night and fasting to dawn is a snap. Then you sleep and from hunger, not a peep. It's just too easy to eat (though not cheap). Eat at night and it's so hard to sleep--but this way (see book title) it's *deep* (no need to count sheep). Now there's no reason to weep over that fat-heap you keep. This may seem steep but if all those hangers-on you clean-sweep you're sure to reap so take the giant leap and you'll have beauty-sleep as you go deep—a must for the black sheep about to rise above the heap.

❦ EASE THE FAST: EAT FAT ❦

55

JUST SKIP DINNER

It's so easy to fast after fat. To stay regular and feel streamlined eat your fruit and fat when the sun is at it's highest. The correct food followed by the daily fast ensures regularity as the eliminative system goes to work in slimming, dilating, opening, water-relieving, energy-producing, pressure-relieving ketosis. Once I saw the benefits of protein/fat alternating with juicy fruit I wanted the whole punch. To fast, nourish with fat for the brain, glands, heart and skin. Starch and sugar wears thin so take it on the chin: you gotta get thin and go within for you're no trash bin and that food-fix is a *deadly* sin. Instead enter the light within-- you'll wear a grin. Stay away from the flour bin for it causes chagrin as you stay a has-been. This isn't spin--just begin to find your twin: your True Self from whom you've been separated (that's why you could never win since you can't-remember-when). Without fat-meals your first fast was famished with forlorn, but eat fat then fast and you'll want to go higher--*fasting longer*: now you're reborn.

ENHANCE IMMUNITY THROUGH FAT

It's when you *stop-eating* that the fat works. As you see it's richly relevant results of staying high in the head—face pulled taut, think a lot, above the rot. You're now a big shot—the best of the lot never thinks "I cannot." That was your blind spot—bad food is like a blood clot: you're *caught* as it ties a knot with you thinking it's so hot. Great Scot--it gave you a pot like a polka-dot. More often than not you forgot that food's a plot trapping you to lots, but fret not: With this new regime you won't walk you'll trot--are you ready for a yacht?

Eat fruit or fat then fast—now the body does as God designed. With fat alternating with the fruitarian diet you'll never be sick or down--no more the hypersensitive suffering with hunger. Since we need protein and fat for immune-enhancement fat with non-sweet fruit is a must (more than salads or vegetables) for those suffering from hyperinsulinism and especially for the recovering anorexic who wants to fast while avoiding institutionalization. The anorexic can stay thin without incarceration by just eating a slice of cheese and figs or raisins and still fasting each day. Now she will be heathy and pretty not bulimic-ugly.

✺ FIND THE LINE ✺

When eating fat you must respect the line between "shine" and "swell." You can fill up on high-water fruits and salads but the whole point of the calorie-dense *potent* foods is small amounts allowing fasting for longer periods. In paleo-science the higher quality (HQ) foods are *calorie dense* while in the vegan (detox) model quality was seen as calorie *sparsity* with the highest vitamineral content. Think this way, you won't be content. We need fats more than we need all that, for we can live on just animal foods (that's eating like a cat). In fact for universal reactors, eating anything else starts combat. With time your fat-deprived system is satisfied, you won't crave fat as much since the system "knows" it gets a daily dosage of the wonderful stuff. It's a matter of respect: you love the HQ, find it satisfying and delicious—but you also know there's a "line" with the "fine": too much isn't benign or divine. Don't be like swine when you sit down to dine. If you wish to shine be like a shrine: stick to God's design. Watch what you combine or see a sure decline. Mixing fauna with starch can malign while eating it alone would refine your features to define you, as thin as a pine— that's God's assign to thine. Eating Neolithically you've an erect spine—the superior sign. Eat your fat but don't forget what's on the vine: the grapecure's a gold mine. For some of you that means wine but watch your line. Yet even if you cross it sleep brings balance and the morning's fast snaps back to home base to win the race (your destiny re-assigned).

JUST SKIP DINNER

It takes a little time to adjust to the new diet but soon you'll know you could never go back to the insufficient-inefficient lowfat vegan menus. For decades I ate that way but there was fatigue, craving, lethargy and even at times uncontrollable anger. After I included fat I was amazed at the highest energy, excitement, enthusiasm and emotional stability from fruit-fat-fasts--sailing smooth in an ageless body. Eating and digestion are the biggest agers and now some fat followed by a no-hunger fast kept me satisfied and energetic much of the day.

If you eat too much and cross the line simply take note and never cross it again. It's just as simple as that: keep telling yourself "I'll never have to feel this way again." To overdo is a terrible drain, a weight-gain, a morbid pain. Are you ready to reign? To stay sane those food-habits must be slain, so now use your brain. It leaves a stain what's going through your vein so choose your hormones and maintain. Will you abstain or be a sewer-main? This is no sales-campaign but a new public domain (half to entertain): will you please constrain in order to attain?

❦ THE SABBATH (FAT) FAST ❦

Due to the rawist publicity and what seems as cosmic fantasy, neophytes excitedly jump right into fruitarianism. But if one has hyperinsulinism the very thing he craves--fructose--is what he's *allergic* to. Even a grape or a raison can degrade if one's hyperinsulinism has come from overindulgence in cakey cultural concoctions. For immature fruitarians insulin-resistance means bloat, constriction and craving along with mood swings, closet binges and fatigue. The bloodstream floods with sugar mixed with released toxins and the result is moodiness, craving and disaster. Avoid fruit-gluttony. If that is your problem and you feel constricted bloat try the week-end fat fast. Have your fat like Italian Tomato Melt, Greek Red Salad or a Spanish Cheese Omelet and delete fruit for the week-end. It's much easier to fast after fat so you'll have the energy for the entire weekend's activities while losing weight. After twenty years of gluttarian-fruitarianism (irresponsible fruit-eating) my Insulin Resistance (IR) was so bad I had to fat-fast for a year (with just one cheese

58

omelet a day) before I could enjoy fruit without constriction and bloat. The IR "level" beyond which problems occur changes constantly--that's a fact. Just restrict carbs and the new life of delicious reversal dieting is soon possible–you'll have pleasures forevermore (packed).

This happy fat-fast will thrill you without the "sore symptoms" of fasting. Watch your skin revive to moistness, your emotions even to sweetness, your vision open to vastness. No more dues to pay—just look forward to Monday. Your pains will allay and fat's gone away. You've got the right of way, so who are they? Now without delay watch that fat dwindle away—you've come a long way. Those vegan starches led you astray (to your dismay). So without delay end decay caused by fat-phobic low fat clichés. Everyday you must scare away that old array of foods which betray. Don't you want to display like a classic ballet? You can go to café (every day) but obey the right way and all else laugh away. Enjoy café au lait but from meals of no-starch fat you must not ☐ stray. Never fear fat--it will not fatten if carbs are deleted and great ecstasy you will reach. Fast after your fatty meal--that's what I teach. Dry skin's gone as it brings water to the cells and surface: now you're moist like a peach.

✂ THE WORTHY WEEK-ENDER ✂

At the end of 60 hours you'll have achieved a new kind of experience—a look and feel you have *won*. Are you ready for such fun? Barring none you must shun all those sugar/starch carbs or you'll weigh a ton. Soon after the fast has begun you'll feel the sun and want to run--the blubber's undone. You'll enter an new reality created by the properly-fueled enlarged encephalized brain of your fat-fed Neolithic ancestors. Become a longhead and return to the fat-fast often after seeing how all water and salt (puffiness) leaves the kidneys, all airways and vessels dilate, the skin becomes ultra-moist and you'll be slim again, amen. Now enter my den (it's a beautiful glen) and try it for days ten. You'll feel so good living among the wise men. After trying so many diets you'll now be thin though full again.

JUST SKIP DINNER

The obese culture deprives itself of fat out of cholesterol-fears. As a result the masses have mass cravings for an infinite variety of foods containing fat—mine were Mexican food, pizza and fried chicken all filled with it. These cravings are specific and people will do anything to satiate them. I never realized it was a *generalized* craving for fat (out of physiological *need*) that created the craving for countless *specific* food delights. For the food-compulsives I can hardly describe the relief you'll feel if you can just grasp this notion. There are only three macronutrients: fat, protein and carbohydrate. Take a little fat and all other fat cravings vanish. Fulfill the *general* need for fat and your *specific* cravings will be gone. The food-compulsive feels "hit" from an outside force like demon-possession over which he has no control. It's a terrible thing and the fat with the cleansing (sweet or non-sweet) fruit diet is the answer. The fat builds, the fruit cleans and both together eliminate craving. Money you'll be saving. On this diet you'll be raving as new paths you'll be paving. This is the only way to face-saving: your face like a steel-engraving and a noble way of behaving.

✥ THE TWO-SPEED FASTING LIFE ✥

The week-end starts our higher two-speed fasting life. I often start mine on Wednesdays or at the latest Thursday for not just fasting but feeling free of tyranny or society. It's a mental trick from childhood when we were happiest on week-ends and this two-speed week is what makes life so much fun. What a relief floating like a leaf! You'll be so happy with disbelief as you fat-fast to be the Indian Chief. Fast and the interim is brief, as quick as a thief you'll have relief. So let me debrief: end your grief by assuming this eternal motif. God designed the Sabbath to be different from the week. Whether feast or fast or the fast *being* the feast, enjoy thy Sabbath.

JUST SKIP DINNER

ᎧGREEK OR INDIAN RED SALADS Ꭷ

Whenever I need to reverse gears from the one meal of apple juice and cheese I eat the Red Salad made with non-sweet fruit and an avocado. Slice red bell peppers, tomatoes and avocado or cheese and dress with lemon, garlic powder and olive oil. You'll so enjoy this: just the deep red color alone ought to tell us something about the incredible benefits on the blood, the supreme ancient varicusa. Nothing else in the universe has the psychedelic rose color of the red bell. It's indescribably delicious, low in carbohydrate (6 grams) while cleansing and rebuilding your rich red blood. Just look at it as you enjoy and feel the cleansing dynamite power (the famous flavanoids in all red fruit). My food colors are *red:* raisins, flame grapes, tomatoes and red salad; *orange*: orange, cantaloupe, cheddar and *yellow*: pineapple, lemon and swiss. *Red* symbolizes fire and power in public. *Yellow* stands for creativity, childlike, mental elevation and absolute world joy. *Orange* signifies world affection, love, humanitarianism and public acceptance of the following ideas:

ᎧEATING CREATES EATING Ꭷ

It is helpful for the compulsive eater to see that eating makes him keep eating. Hunger is incurred by eating the wrong foods, the deletion of which takes away all hunger. I was especially amazed to learn how I had no hunger for days after fat like doughless pizza. Why does eating (the wrong food) keep one eating? Because as soon as one stops eating withdrawal sets in as the stomach cleans out and this is the painful "hunger pain" which subsides with food. Like an alcoholic we *eat to avoid withdrawal.* The result: a gut filled with fermenting food that never really empties. The one fat meal followed by fasting will preclude all this because the fat-fed gut is comfortable all day while fat-burning, as all excess water (bloat) is expelled through glucagons-elevation. By fasting in the afternoons the assimilative and eliminative system takes over from the digestive—we eliminate a neat package the next morning like clockwork. This is exactly as it should be and eating dinner blocks

61

JUST SKIP DINNER

this wonderful process. Whatever you eat or don't eat today, do it only for *how you'll feel tomorrow.*

One eats because he eats. You must avoid all the treats especially the (non-fruit) sweets for they'll bring defeat worse than the streets. You must let reserves *deplete* to release HGH—the purifier-youthifier making you petite elites. So let me repeat: take a retreat to thin-out your seats for it's the only way to compete. Then no more they'll mistreat with all their deceit. Are you ready your work to complete and to change all whom you meet?

✺ EATING IS NOT TRUE PLEASURE ✺

Another idea is that eating is seen as pleasure but it usually isn't. I told one lady I had a small potent fat lunch and then just skipped dinner. She was in disbelief: "Don't you realize the pleasures of *eating*?" Pleasure? I remember the years of agony, craving, constipation, feeling stuffed, fantasizing or fixing food, always running to the store with a "whim" and constantly cleaning the kitchen. This is pleasure? Add to that the misery of getting fat and prematurely aging like the rest of unconcerned humanity. Most all this misery comes from food (but also mal-adapting to the rude) yet it's always given another disease label—it's one big fable. Don't mislabel that food on your table. It makes you unstable with a big sign "not able"—and another "incorrigible". But fruit, fat and the fast turns you on like a jumper cable. This is a blissful life from energy's higher pleasures but these sensual enjoyments are just icing on the cake. The real rewards are True Selfhood (deep roots) and your work coming out—bring out the flutes! You will now find your destiny, purpose and genius—that peculiar bent's worth much more than a cent.

✺ NO NEED FOR GREENER PASTURES ✺

Another idea is that there is no more need for endless seeking of greener pastures: buying boats and vans, going on vacation or to

JUST SKIP DINNER

restaurants, playing mind games with people or getting lost in worldly entertainment. We're always told of the glorious places to go: Vegas, Florida, the Riviera—the throng. Wrong. The most glamorous exciting place is one's own home in the fasting state. Yes—that is it! We are the chosen few who know this new but ancient information. The fast will cut through all you pursue as the untrue you'll see-through. Change your world view. Become strong and shiny like bamboo and you'll have it all beaucoup. Now stick to this plan: don't let it fall through. Become deaf to those luring you to eat wrong. Are you ready to cling to the truth and to the false, eschew? Now ensue to the plan so beautiful fat goes to the skin not the thighs: let it soak in to remove wrinkles (imbue). You won't be blue as the world now sees who's who. To dietary fat say "I do." and to the fast say "I will make do."

❧ SMALL GUT LESS ELIMINATION ❧

The fourth idea is that as fruit-fat-fastarians the stomach shrinks to the size of a walnut and eventually becomes a mere appendage along with the intestines and colon. With such a tiny but powerful diet things become far more aesthetic and suitable for nobility—ladies and gentlemen. American's are constipated by constantly using their gut and intestines for illogical (to the body) substances. With fruit and colon-happy fat the colon's function is neater, niftier and totally regular. Who should worry more about their health: those whose primary organ is the gut and sewer or those staying in the higher brain above the bulk and bog? Stay high you'll be so happy you'll jog. No more eating like a hog and you'll sleep like a log. It's like coming out of a fog.

5
FRUIT-FASTING

Fruitarian bliss is God's kiss and something you won't want to miss.
Without this recurrent cleanse life's amiss. But when you feel
constricted or bloated just switch to swiss.

Reversals between fruit (Ehret) and fat-fasting (Atkins) combined with daily fasting is a happy, healthy and fulfilled life. Temporary fruitarianism is a very high experience: some days are fruit-fasting vacations like a few figs or the grapecure for highest beauty, ethereal reality and healing. Red Salad and other non-sweet fruits would also be a fruit-fast. Fat-fasting follows this wonderful phase with rebuilding, appetite-suppression, fat-burning, energy, stable moods, clear thinking and moist skin. Reversal dieting between fruit-only and fat-fasting with non-sweet fruit and leaves is a varied and fascinating two-speed life. "Fasting" can also be defined as: "just eating less".

❦ EAT THEN STOP-EATING ❦

This is the secret you've been looking for: the higher delights of not-eating or just the *intention* to fast. Even though you may end up eating just start again for it's your mere *plan to fast* that evokes PHF: positive healing forces that flood every atom of your being! There's no higher high than fasting or intending to *not eat*.

64

JUST SKIP DINNER

Just start and miracles work everything out to perfection: Lost things found, enemies overcome, revelations for success, miracles, lost fears of aging, all problems resolved and solutions illuminated—that is the joy of fasting. Whenever you feel down and out simply intend to fast and presto your miracle arrives. Do it a few times and you'll be addicted for life as you say good-bye to strife. The fast cuts through problems like a knife.

SMOOTH AND SAVORY SAGACITY

The fast opens the door to a new perception of reality—a clarity you've only dreamed of in fairy tales. Become a "discoverer" as inner and outer worlds illuminate as a kaleidoscope. In the fruit-fast even walking in 125 degree heat is exhilarating. There is no fatigue, just excitement as the boundaries between self and universe (sun) break down and one becomes the sun—then the solutions to all problems and answers to all prayers illuminate in consciousness. "White" society (mucus-filled and dense) is divorced from nature, the white man never fasts and the white woman worries over wrinkles. With your reversal into fruit or fat-fasting all wrinkles and worries smooth out to calm perfection. It's so calming and blissful you'll have great affection, divine connection, a new direction, angelic protection and beau-reflection. Opportunities are your selection for you're in the winner's section.

CLEANSING RELIEVES HUNGER

Why do people crave to eat? Either beause they need to clean or they're eating wrong foods: the toxic allergen excites hunger as eating veils the poison. Learning how this works starts your new life of living without that old hunger. Just eat right then stop-eating as you fly off to bliss. Make a serious new dent in your clan or clique today. Eat your one major meal of potent fruit or fat then fast into perfection: this trick is a fat-ejection, prevention of infection, a soul resurrection and a foe-detection.

JUST SKIP DINNER

Both fruit or fat-fasting leads to *encephalization*—enlargement of the brain and skull--whereas huge vegetable salads, starches and grains diminish the brain and bog the whole process of digestion down--killing the crown, wearing a frown (hardly renown.) From many big meals one will drown. Only one major meal is needed (or continuous fasting *punctuated* by mouse-meals like one nut, fig or slice). It is a matter of brain-gut competition of energy: many meals require so much energy that the brain is robbed and dragged down to the gut. Eat one big meal and the colon works on the principal of *mass evacuation* as the digestive-assimilative-eliminative systems sequence throughout the day.

Once you digest you're back up in the head not feeling like lead. A little fruit or fat releases energy from the gut to the brain. Clearer, sharper thinking and miracles are the result. Once you begin a life of fruit-fat-fasting the energy explodes. You leave the compendious eating life behind and join the high ranks of genius throughout the ages who found a new and higher life through tiny but potent meals followed by not-eating (paleo-fasting). Food gives one a beating. It's a kind of cheating your True Self a-fleeting, no way of competing (it's only defeating). But fast—your work is now completing.

⬥ FRUIT AND THE GRAPECURE ⬥

The modern vegan fills up on salads. The problem here is that man is not *folivorous*—a leaf-eater or plant-eater. Instinctively (if free of culture cravings) man should mistrust limp soggy vegetables and starchy spud. The fruitarianesque higher paleo "look" is sleek and sinewy, lithe and lustrous. That shine won't come from spuds and veggies resulting in the "bready" look as skin turns to leather filled with white sand. Whether veg or spud miracle-consciousness is banished by these foods for lower animals. They have to eat too but man in his lower ego consciousness thinks everything on earth is for him! Man's *higher* reality illuminates from fruit including dried fruit like organic figs and skin-moistening mood-elevating fat like avocados, nuts or a little piece of cheese (fauna-fat). Whenever I've eaten cooked produce or starch it was a great relief to return to nuts, cheese and apple juice to influence the body machine for the day's activities. It means eating very little for all-day eating makes the bones and skin dry and brittle.

❧ STUFFED WITH STARCH ❧

The fruit-fatarian-fastarian is like a happy, shiny, slim child. What you eat can make life a defeat or wonderful treat. Starch fills the skin with "land mines" of deposits as it destroys higher fastarian consciousness. After a day of starch you'll think "what happened to all my higher thoughts and great optimism" or "what happened to God—I feel like a clod" or "why am I so itchy (bitchy)?" After a small fast brushing your arms and legs will reveal white spots as the "mines" are brushed out and flakes fly through the air. That starch is bad—no wonder it drives immunity mad and makes your mood so sad. Now return to just fruit a little fat and your daily fast: You've returned to joyful thoughts at last. These episodes of trial-and-error make you the best so just regard them as your test.

❧ QUICK PERFECTION: FASTING SELFIX ❧

A refugee from fruitless fruitarianism and orthorexia I felt so much better realizing I could eat out and enjoy fish and salads or when home an ultra-delicious dainty like "dough-less pizza" made by melting cheese and parmesean with tomato paste (or sliced tomatoes) loaded with garlic, cashews and unlimited olive oil. What freedom after such bedlam! Even with deviation there's a swift return to perfection by daily fasting creating the look and feel of polished stone and God-revelation full-blown. It's the fat then the fast that's the blast. Then you're the highness as the fat-fed brain evokes hyper-creativity all day long: putting things into place while the old life is gone, ne'er a trace. Watch your face become prettily geometric

JUST SKIP DINNER

like lace while you lose interest in taste and the complexion becomes red like a new race. With the cravings finally gone you're no more a tiresome clone—you've got your own magnificent life and how it has shone!

An example of a transitional fat-frugivorous-fastarian diet would be to begin the day with your usual morning drink and a shot of grape juice. For brunch have peanut-butter, Greek Red Salad, Doughless Pizza or fish salad. Now enjoy the digestive process all afternoon-- you're not hungry and you're very happy. There will be periods in your life when only fruit-fasting is indicated but don't deprive yourself of the bliss of balance from fat--for doggedly sticking to diet dogma despite disaster is destructive. Read your body's needs. I know that if my skin feels itchy or if I feel distended or nervous it's time to switch to fat-fasting. Sometimes just some macadamias for lunch will suffice or occasionally I'll begin a 60-hour week-end fast with dough-less pizza then go into this needed and wonderful rest. These healthy reversals make life the best! Just one fasting aft can put you on your crest.

✍ CERTIFIED ORGANIC DRIED FIGS ✍

The black fruit is amazing and efficient. Few of us can get all-organic produce (who wants to pay three dollars a tomato?) so by buying certified organic dried figs through the mail you know that at least your breakfast and the early half of the day is organically pure. That's half the day with the fig which is the fastest cleanser (rate 30): twice as fast as the raisin which is twice as fast as the grape—you dig? Black dried fruit is two to four times more cleansing than juicy. They are easily stored out of the refrigerator for months--a real bonus for the recluse in nature. Have some figs for breakfast and then later some nuts, a spoon of nutbutter or piece of cheese. Ok you've eaten now fast until the main meal or for the day. If hungry at night eat a few raisons (which are actually the grapecure for cancer as well as the Queen of the Fruits), nuts or a piece of cheese. The grapecure is

JUST SKIP DINNER

simple as raisons are easily stored for weeks or months. The black fruits are perfect mainstays—"staffs" for the faster to lean on. If the sugar (fructose) bothers you just stick to the non-sweet Red Salad, greens with guacamole and nuts. If you're a universal reactor, stick with fauna (just cold cuts for sensitive guts). As the fat and fiber greases the runway you'll be pleasantly surprised at the next day's elimination. Later when entirely clean you'll be living like a hummingbird--on a little fruit and a bit of fat: you'll like it like that in fastarian bliss. Without it psyche's an abyss, we're often amiss, careless, remiss. Fast and enjoy the here an now--no need to reminisce.

Fruitarians: unless you live on a fruit farm you're always running to the store to buy fresh fruit. That makes no sense as it dilutes and interrupts the inner journey. Buy black fruit or PB that stores for months—enabling you to fast without fear of spoilage. The greatness of the black fruit is how it eliminates and dissolves "hunger pains." Eat one fig--you're only eating to extend the fast so little is consumed. I found juicy fruit way too water-logging and filling. I like calorie-density, the High Quality Intensity: just a few figs or raisins suffice to do the job of nourishing and cleansing without expanding the gut—that old rut can finally be cut.

⋘ FRUGALITY IS A FULFILLED HOUSEHOLD ⋘

For the entire period of my inner journey I had a very well-supplied household. Miracles never ceased as my food staff increased--I always had my staff to lean on in the form of figs, raisins, nuts and PB; canned fish, olives, tomato paste and pineapple chunks; bottled lemon juice and olive oil. In my wilderness period I have seen all my needs fulfilled by attraction alone and this has educated me to drop worry and fear as long as I'm daily fasting. When I've slipped I always return to perfection by this divine food-election. Having collapsed the whole eating dimension life has become so carefree—and also fail-free since a faster lives in daily miracles solely for his benefit. The stored fruit, nuts and oil keeps me from starving so in gratitude and a desire to extend it longer I train myself to eat just enough when hungry and shrink the stomach to the size of a walnut (yet never feeling "deprived" since I have a year's storage). It has been so great rarely having to shop as I in my new "found" time have gone from the bottom to the top: distractions released I'm ever in my prime. You'll be as powerful as King Tut if you avoid the eating rut: just the constant food thoughts darken the aura like smut while you resemble a mutt. Enjoy a "food time" each day of about four hours: have this be your grazing period and when it's done it's done—enjoy the sun. Switch desire from food-fix to fun: soon you won't walk, you'll run.

AGE DOES NOT MEAN UGLY

Most people equate getting older with getting wrinkled, weighty, wasted. Not true. Getting that old look comes from the wrong diet filled with starch and other fat-phobic substitutions. An older person on the correct fat-frugivorous diet maintains the beauty of the True Self which is God's design living on the food He *designed* man to eat—the fruits and fats which keep the skin moist and wrinkle-free, the energy high and the brain alert until the last day of life. The fat-fruit-faster looks ahead with glee as he continues to become more a part of eternity—for everyone around he's "delicious" scenery. The whole culture looks older earlier, for not only is there not enough fasting or fruit but fat has been substituted with modern agricultural (not paleo) foods such as grains, beans, rice, non-fruit sugars and pastas. The skin system loses logic—elasticity--just as it becomes filled with tissue debris: white chalk caulking every pore, and it is tragic. This life's a bore and digestion's a chore so close that old door. Grains galore make you just want more so I implore you to end this old war then come to the fore.

DAILY FASTARIAN

Now that you know what to eat the most important thing is to fast—daily. Every day go as long as you can between meals or even better: see your life as a continuous fast only punctuated by fruit or fat mouse-meals. I fast daily and in the afternoons I feel so blissful knowing the joy and sense of accomplishment that will follow the next morning. If you used to be an afternoon-binger this sense of achievement, purity and newfound self-esteem will be doubly rewarding. The next-best way is to eat two meals six hours apart and then fast 18 hours. Eat fruit when hungry in the morning and follow later with some fat. Now you'll be calm and crave-free as you fast and thrilled when you see the vast value of fasting—in your appearance, energy, outlook, creativity and definitely your prospects for the future.

If you remain an eating fool you'll get old-looking with the rest. If you eat right you'll be ageless and more alluring with the years.   In a cruel ageist society that looks down on older people you're about to show them the true score! Unlike them you're forever a gem to the core. All those ageist-comments will now be of none effect or will cease altogether for you have now transcended the aging problem (it's a war). You're a lot more handsome than youth stuffed with starch and more.

❧ THE TRUE FEAST ❧
Is not Filling the Gut

In this matrix the "true feast" is not filling the gut but--in order to stay high--eating the highest quality food and thus needing only one meal or continuous fasting punctuated by tiny but potent portions. Eating tiny amounts allows us to stay in the fasting state, i.e. high in the head. Keep it tiny, you'll be shiny. This plan gives the best food on earth for man while also getting the sensuous (beyond all imagination) pleasures of fasting. As fruit-fat-fasters we are separated from the herd because (1) they eat the wrong food (grains, cooked or fast food) and (2) they hate the word "fasting." The incredible pleasures of stepping into the galaxy of fasting consciousness is something few will ever know. Instead they fill up on gargantuan salads calling it "healthy", not tiny meals bringing brain enlargement—the "longhead" marking the monarch on his magical Inner Journey.

You'll have periods when you need a change from fructose (carb) to the fat-fast—this hormonal switch changes every atom and gets things going. Just delete sweet fruit and extend non-sweet fruits,

greens and fats. Soon one has better looks and energy than ever as fat means "spirit" in Greek: it acts as an electrical conduit of spirit as we conduct the source. The weekend Fat Fast makes you a live-wire as does cutting off food by noon to an eighteen-hour fast to the next morning. Done daily the bounty is cumulative and *nonsummative*—the whole will be greater than the sum of its parts. This life is like fine arts as pure bliss it imparts: energy off the charts, soothing sore hearts and perfecting all parts as soon as it starts.

◈ GREEN WITH ENVY ◈

Most health enthusiasts eat a lot of salads but man is not *folivorous* (plant or leaf-eater) for this requires bacterial putrefaction in the big guts you see in huge *folivores* like cows. It's too much trouble and fixing for me, but many prefer leaves finding that the bitter balances the sweet and it fits well into dining society. That's OK for the main point is fasting after fat, and man like the fauna-frugivorous ape *can* eat leaves (they're just energy-thieves). Fat-Fruitarians may eat 60% raw fruit, 40% fat--choose your own ratio even 100% fat like a cat. Just lighten your diet to speed up your life to wear a top hat. The balanced fat-fruitarian becomes as busy as bees in a life of ease but starch and grains irritates like fleas. These are the keys so get down on your knees and thank God for fauna and fruit trees for with foods such as these you'll never see disease and heal in degrees your psyche to appease. No more people to please for just *this* moment you'll seize. It is insulin-elevating starchy grains that makes you sneeze even wheeze, that's my expertise. So on truth's behalf   let fruit and fat be your staff—such joy makes you laugh with energy off the graph while cutting your weight in half.

◈ NON-SWEET FRUITS AND COOKED FAT ◈

JUST SKIP DINNER

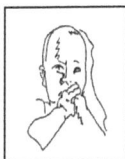

Besides sweet fruit keep non-sweet fruits like tomato, eggplant or zucchini with fats for your main meal. These will be your transition rather than Ehret's advice to use salads then streamed veggies and spuds. So much eating—vegetable and spud is soggy making you groggy. My legs bloated up like tree trunks on his transition diet and the skin became like scratchy cardboard. It always feels so great relieved of the effects of starch. Non-sweet fruit and fat—that's where it's at. A few raisins or figs in the morning is fine but more sweet fruit cleans too fast and the eliminative organs can be overwhelmed (cooked fat slows it down). Even Ehret stresses that going all-raw is an overly aggressive mistake. Man has been cooking from the paleo beginning: we've adapted and according to Billings cooked food is even more digestible ("pre-digested") thus releasing energy to the encephalization of the brain. The toxins are cooked out and the vitaminerals are more bioavailable. A delicious satisfying cooked meal like "tomato melt" with olive oil, garlic and Italian seasonings (incredibly healthful) tomatoes, zucchini or eggplant with cheese is perfect for your one meal with energy for the day and no hunger to pay. But if starchy spud is used in the transition it is stored in the skin and lymph--hypersensitive-ectomorphs can look like the white polar bear not the angelic nymph.

RECAP:

❧ SWEET: NO-TEETH, STARCH: IMMUNE-THIEF ❧
Fat Makes Smooth and Shiny—What a Relief.

Like many fruitarians eating fruit all day I lost my teeth and suffered hyperinsulinism, exhaustion, mood-swings and cravings but after a two-year fauna fat-fast followed by this FFF plan I stabilized into a new wonderful life. With the new calcium my new (restored) teeth stayed healthy. The lacto (cheese) and nuts will slow aggressive cleansing down or in your own sludge you will drown. Your health will improve as you gradually clean and now you'll have sheen. The energy from fat is keen and you'll look the best you've ever seen.

❧ FAT IS CLEANSER—IT'S YOUR FRIEND SIR ❧

73

JUST SKIP DINNER

Many people do very well on nuts and lacto-fruitarians on cheese for they are high in fat and man needs that. But Ehretists will say they are not cleansing foods like fruits and vegetables. However by elevating glucagons, cheese is cleansing in another way (fat-burning, nerve-dilating, immuno-enhancing, energy-releasing, appetite-suppressing) and the nuts are maximum colon-happy eliminators. I am amazed at my new level of health with the fruit-fat-fast and also how the body has streamlined without any more unexplainable "fat pockets"— clothes fit perfectly but looser now. Living on raisins, nuts (a most delicious combination) and cheese results in miraculous regularity each morning. I am thin and quick with no fears of aging (watching the clock tick).

It's not saturated animal fat which causes cancer—it's vegetable oils cooked at high heat. You will lose health from no animal fats, for we desperately need B12 gained from no other way. It is our own liver which creates dangerous cholesterol, cued by insulin-elevation from eating starch. In contrast, eating cholesterol-rich foods will *burn out* the dangerous cholesterol instead. Now its all explained why you so-gained. You were un-trained but now info's retained. You complained (were so drained): to the grains you were chained and after so many diets you regained (for Jenny Craig you campaigned). It was all so ingrained so from wonderful fat you abstained (so more fat you obtained.) Then hunger became unrestrained: it was all so unexplained your patience became strained. If you ate wrong your talents were stained though your success was preordained. Well now you've ascertained the truth , so energy's sustained.

6
STRANGE LIFE
Fasting Revelations of a Fruit-Fat-Fasterian Chieftess

Life is strange. The faster has a psychic knowing about what's coming as the seasons change. He just knows and in this way stays "on top" of things. Déjà vu? Clairvoyance? Precognition? He knows it all but he matures when he finally learns to *trust* these instincts without recourse to "reason" or habits to avoid anxiety. When you fast--on people or habits—just let your mind go and trust your guts while avoiding all ruts. I'll tell you the whats: you've been dealing with nuts for when it comes to sick systems people are mutts. Just like in a dogpack the Black Sheep gets cold cuts—he gets the "shuts": he is shunned, labeled, devalued, invalidated and kicked in the butts (treated like a klutz). This life is "awfully" strange when stuck in sick systems but for the same poor soul it becomes fascinating as a faster—that's the kind of "strange" we want. It's the difference between culture and crudity; elegance and abject poverty; popularity, prosperity and fame vs. humiliation, rejection and shame. Fasting—on people, habit and food—is the answer for all problems even psychosis. Whereas she was a mute now she's the hostess with the mostess. It changes all prognosis, it frees from sick symbiosis and on all business and social relations it spreads God's hypnosis.

Yes life is fantastically and beautifully strange as a faster because here *we're* the master with God as our Pastor. Steeped in

JUST SKIP DINNER

synchronicity—how everything fits together in a jigsaw puzzle—the happy faster lives in a fairyland. High in the head it's one miracle after another. Clear, he's psychic as he sits in the collapsed moment where past and future unite. Amazing magic coincidences make each day exhilarating. Enjoy this day with me as you journey through your mind: you will see, you will find--all else is a bind. People can be blind and so unkind—a sick system is a grind (you should watch what you sign). But eat fat and fruit then fast, and it's all left behind—you'll have become refined. Those others with whom your sins intertwined maligned you as they debased your state of mind. This is a story of mankind so become resigned. Will you remain confined or be more inclined to take the position you're *destiny* has assigned? This one is aligned with a superior force— through you it's always shined. So let me remind: through obstructive sins you declined that Great Reward, and as you wined and dined you mis-combined and your destiny was declined. So in this essay you'll become new (you'll be redefined).

⊰ ENTRAPPING SITUATIONS ⊱
Split from Mundaneity

Split--from liars, idolaters (people-worshippers), thieves and passive-aggressive vacillators--to a new culture or race and the brain explodes into new dimensions. You're too mind-set. Make your mundane world absurd to your great relief: then the faster *is* the world. Are you ready? Listen up: many of you were rejected and scorned by family, friends, society. It hurt until you learned to concentrate on the highest so that all things *below* could go strangely dim. Damn them? No—forgive all. Life's too short: you want to make a mark instead. Focus on the full moon, absorb the sun, fast, pray, love those who love you and forget the rest. That's the test. Resentment kills: it makes one ugly and mean. Don't you want the sheen? I mean flourishing in your gene, staying keen. You'll be King or Queen if seeing through this scene as the system dissolves—then perfects—like a machine. Look like a teen--so lean, so clean (just give up the bean).

JUST SKIP DINNER

Once one's entrenched in sin the groundwork is laid for the sudden emergence of group grudges (the group mind or "collective unconscious") against the sinner. Then a seemingly trivial event occurs and he is pegged, perhaps even locked up. Then the agency has all the witnesses they need--the trivial crime is blown up followed by punishment for *cumulative* sin. In families the same occurs. The family is history, so the trivial is magnified by the group mind: they fill their cup with one insignificant event as they pull him down. Now the good or bad effects of these incendiary group grudges are completely determined by whether one is eating wrong or fasting. Any scapegoat eventually learns what he needs so the scapegoating will cease: he just fasts at the scary group routine which all dissolves like a vapor.

Be beautiful, love and forgive all. Most importantly, be nice though you see them as lice. Don't insult, just reject—that's the elect: maintain *your* dignity in the face of people possessed by cruelty. It's the nature of these times as most come from dysfunctional families--but through these rhymes we'll dissect the *essence* of these crimes. The tendency is to fight back, bear the brunt of another's hatred or become sick over it. If we can transmute that energy—transform the fight into a fruit— we'll have won the greatest battle in our lives and make gold. It isn't easy. When a whole system hates one member (a sign of the Great) he's been given the golden opportunity of *assured success soon* through the fine art of energy transformation from the worst to the best. Do this and you'll be blessed. Open your chest, succor the breast of mother nature by *not reacting* so the energy will explode to new dimensions--that's the crest. This is no jest: just be a knight tonight and come back to your spiritual nest: Now, no matter what you've done in the past you're the divine guest. Here I suggest how to deal with the pest and all the rest: You've been severely stressed (it was a giant test) but if you can see it as a contest and finally win you'll have real zest. I can attest: I was depressed as my soul abscessed--it made me obsessed (just by how they dressed I was impressed). Read these words and you'll be ready to fly not buy the lie and die.

☙ FASTING: SPRING TO THE PALATIAL ☙

Through fasting we spring to the *palatial*: of or relating to a palace suitable to a magnificent mansion. The word "palatine" is like a holy or Roman emperor: the exemplar of royalty. A feudal Lord issuing sovereign power over and within his domains. Yes this takes brains but *mostly* the cutting of chains causing pains. Only this can put fire in your veins in a land where it mostly rains. All you've known is hills, valleys and rocky terrains--this system explains your morbid remains. But seeing through all this will give you the reins: your gifts retained the King now reigns.

What were these chains? The sick system was a *net*: an entrapping situation, a group of communication stations operating under one unified control—the system set-up, the status quo, the group mind. Though just one evil helper could've kept you in a bind, it was still a *network*: a fabric or structure of cords or wires crossing at regular intervals and knotted (glued). These were all the supposed needs this dark knight helped you get, or the interlocking jealousy patterns and secret alliances of two or more against one—the forces making you mute or psychotic. It sickened us at our source as we became *neurogenic*—disordered from abnormally altered *neural* relations. Now that you understand your irritation, know this to win: be *neutral* for immunity from the invasion of belligerents. Stay centered in yourself and refuse any more tangents. Don't join, call out or attach from fear. Now you'll become *newly*: new and fresh in a new way. This always happens when you break out of a *nexus*: connection link or connected group. If you can transcend this sick system keeping you down it will be like *Niagara*: an overwhelming

JUST SKIP DINNER

flood or torrent of blessings and you'll become a *nicety*—an elegant or civilized feature with Great Power in this world. This inversion will occur every time when you detach and return to center. Remember always: *detach, return to center*--for God's spirit lives in you and will not re-enslave you to any more fear or supercilious people.

⚜ ENSLAVED TO DECAY ⚜

You have been *enslaved to decay*. This means slavery to other men: bondage. Through the net these form cobweb-illusions in the mind holding you back. It can be a young gang member or a "kindly" old lady—don't kid yourself, anyone can be shady. You must get into your own stream and *now*. The fast separates you unto the highest calling, the treasure hard (impossible) to attain through any other method, and the last crisis we will call your *separation unto your highest calling event* after which you will never experience this pain again. Now it's welcome to elation: you're in a new station, my friend. We gain access to the highest through the fast which is a secret tunnel. We then become *pathfinders*: those who discover a way—a prophet, an inventor, a statesman (that's you today).

The sick system was a severe drain, an *eclipse* as you were obscured by another body bringing decline, disgrace, obscurity. It was a question of *ecology*: the relational patterns (in totality) of the organism-to-environment. This brought *ecotone*: tension between two competing communicants. Then you the victim fell down and abused habit to avoid anxiety by becoming *edacious*: voracious in eating, drinking and other devices to soothe your broken spirit. Then you (the black sheep) became feeble and decayed because of a possible *edict*: a public proclamation having the force of law. In other cases the victim was *effaced*: wiped out, obliterated, made indistinct by rubbing out. In the eyes of these others (and later yourself) you may have become an *–een:* a small, petty or contemptible one. After all this you (the depressant)

79

JUST SKIP DINNER

becomes *eerie*: cowardly, wretched, frightened, strange and gloomy. Such is the evolution of sick cycles and systems.

But now through these pages you'll be educated, making you stronger so you can transcend the mess making you less. Now this will *educe*: bring out something potential or latent. This brings comfort, for having released the obstruction all things can now work together for good. Since in this system we got weak (sinning to avoid anxiety from persecution) we were cast out from our rich destiny. So now just repent and the sick system dissolves. Fast so your detractors become suddenly ashamed and you'll take the throne, an ex-clone.

✺ SUN, MOON AND STARS ✺

Sun moon and stars. Keep these central—all else peripheral. Always return to this source (the elemental reality) when things go wrong. Thus you'll avoid the trivial—the petty irritations of everyday life. God made the elements so appreciate these things and you'll know God, the Almighty Giver of energy and health, for He's the way to wealth and stealth. The more fascinated you are with the elements the more expanded your vision becomes.

This is bliss—welcome!

Now look out to eternity—trust your instincts and every single idea that "comes up". That way you're rich before you're rich. You're only problem is *not* pursuing your fasting ideas. Trust each one and go for it—don't forfeit nor forget these dreams of yours. Following these instincts won't be chores for they're magic doors and peaceful shores. As the eagle soars your talents they cannot ignore. So try a day's fast and go out of doors exposing your pores. You'll find your mind loves to explore, then lay down and let God heal you gently. You will win out over the sick system if you sanctify this day: fast and pray. Count your blessings and forgive your messings. Fast and you are protected under God's giant wing so just sleep and go invisible—not a peep.

✂ SPLIT FROM THAT HERD ✂

Follow this plan and split from that herd. It hurts—I know, I was there. I was rare but they didn't care and I didn't dare. You've got flair (as God's heir there's no compare). But those others can really scare, so beware and prepare. Take a dare: live on air. No more you'll wear the shape of a pear with tires to spare. That food life does nothing but impair, like a dark illicit affair. But through fasting you get clear and realize how different you've become and now alone you are sanctified, separate, soaring. It'll never be boring when you no more look like Goehring. But with talent and love out-pouring and the public so adoring no more they'll be ignoring. So now I am underscoring that eagles soar alone but chickens stay in coops. Stand proud eagle, as the True Self enthroned in nature.

✂ EVERY DAY SATURDAY ✂

Fasting is all about consciousness—freedom. Think: Saturday—that's when we felt free right? We could escape the mind-track and flee to fantasy and fun. This fantasiacal escape must be our new daily routine. Free from the blight, society's flee-bite--it was a fright, that fight. The system was so tight, while we were always contrite. It was an awful plight as they took such delight in bringing down our might—how they loved to excite (each other) to ignite. This is an understatement (I'm just being polite). But release these systems and cycles and soon you'll be at your height quite

JUST SKIP DINNER

out-of-sight: the Queen of the Night. It's like a religious rite the fast (taking flight to the bright).

This bright life is the elemental reality and the two-speed week. Learn to live it: mentally mine starts on Wednesday. This is when the mind can wonder to eternity and come up with winning insights--the solutions to all matters, having released the stress from the week and it's chatterers. I usually take a five-day week-end which I intend as deep rest being in another reality altogether. But because with relaxation the creative insight erupts these week-ends turn out to be a frenzy of creative work activity--"putting things into place". In this space one cannot stop for with the creative pop you'll finally get to the top.

Before a creative artist is rich no one takes him seriously, so to finally see this light fills him with might. He patiently makes sure it's all right then at the end of the long haul genius must fast to hear the call. This is the time to rest your gut like none other for if you eat your talents will smother, but fast and you'll *finally* get your new money-mother.

⊷ FASTERS MAY MESS BUT NONE FOR LESS ⊷

We all make crazy mistakes but the faster gets the constant gold stamp of approval. His enemies are amazed for the guy cannot be kept down. The faster is a magic mystery as he maneuvers out of misfortune and makes miracles instead. Many suffer constantly and die of broken hearts (called "cancer"), stroke, heart disease. No one understands how these world maladies start with food or why so many end in "depression". Learn the precious lesson: one's spirit is stuck down in matter--obstruction from people, habits or food

JUST SKIP DINNER

(messin'). But if you learn about fasting—the fun of fantasiacal transcension above subservience, reversing the matrix to our *dominance* in this demanding world, you will win. People eat to deal with the dregs in their situation. Eating never works and fasting always works. It's the lesson I came here to learn and teach: The world's like a leech, a trauma for each. This makes it hard for people to reach the implications of this speech. Now I don't want to preach but always recall the importance of fat (not just the peach) and then fasting to overcome the breach. So eat right then fast, I impeach.

Lost a friend, job or plan? No worry it'll all be replaced with something far better than. Such is the path of the faster: constant changes. For he lives in the sky high above the disruptions, distractions and carnal desires of ordinary men.

Take joy for now it all starts to get better: there is nothing to fear for the future is fullness after removing the fetter. It just takes knowledge and maturity leading to insulation from harm and this is "enlightenment." Incautious involvements mark the immature for those user-"friends" really take their tolls. Neurosis is getting stuck while maturity is freedom and *increase into more meaningful wholes*. Are you ready to achieve your goals? Get clear lest you be like the Dead Sea Scrolls. End all controls: these people never extol you for they only "think" through the polls--they guard through word-patrols. They're filled with holes while you've got *goals*.

✂ ALL RECOVERY IS ELIMINATION ✂

Pray for success and then you may meet a crisis, for God changes by *chopping*. He either prunes your life or chops superfluity off suddenly so hold on: after elimination is complete, your success is sure. Remember the formula: all recovery is elimination. Fastarianism can be a lonely life— vastly different from the ordinary mass eater-evacuator, as we are the few and far between. It works well if alone but there is great suffering on the work force or in a sick

83

system. I am sorry for your sufferings, rejections, insults, being put on the defensive and the need to explain. But soon you'll be on top and experience constant successes without refrain. So persevere my dears, the end is near so not a tear nor want of beer. You've got flair, but do they care? No, they just glare so you feel you haven't a prayer. To be sided against is a scare. Being rare we all share in this event of great despair. When you become aware you start to beware and out of this comes strength of no compare. So I declare: you'll no more be a square, just a multi-millionaire.

❧ PRESSURE MAKES US FASTARIANS ❧

It is *pressure* putting us on this solitary path. Since everyone seems to be on the wider path the inner journey gets lonely. With me it was fear of chaos that brought a longing for order (the basis of Puritanism) and thus solitude. But if we can persevere the lonely but neat path culminates in joy and success. Ah the feeling of finally going beyond a problem: that feeling of transcendence as it all dissolves is the greatest event: Have a good cry then view the mountain vistas. For women books on feminism and the plight of females can heighten feelings of vulnerability. The answer for them is the same for men: just fast and pray daily and get to know God.

When your life gets cluttered with the non-essentials just return to the elements: sun, moon and stars. Expand your mind—just think about Mars. This is much better than cities filled with cars or idolizing superstars. We've been behind bars--it was a crisis of ours and we show the scars: being belittled by czars (braggarts smoking cigars). But fast and out comes the guitars! The female would do well to be sweet, pretty and neat. Never *demand* respect—just *command* it through true character and utter purity. That's the beauty seat on Success Street. When it's insults and anger you meet, don't fight back but kindly greet then walk away gently—that's the strength of a fleet and they'll remember you as sweet. That's the way to compete: think what you want of the indiscrete, but keep *their* memories neat.

✂ TRUE (ELEMENTAL) REALITY ✂

Early on I learned about reality: that we either create it ourselves or become socially-hypnotized to tow the line. Like most hypersensitives I felt an outlaw. As others sensed I could/would not conform I felt ignored and desperately lonely. I know how cruel conformists can be as they sense a difference. You must gain strength to be accepted in your own right. You've got that right: to learn how to do it is your plight. It won't take you decades if you learn the solution: never get discouraged and have faith, for once your work is complete you must *wait*. It is seed (you plant), time (you wait) and then the harvest (you celebrate). Wait, knowing your reward will occur. It's faith that is needed lest you forfeit all in frustration. Just fast and keep climbing and shining: desire the topmost (quest for the best) and upgrade regularly. Keep leaving lower levels to reach higher ones. Mean? No we're talking the sheen of a Queen. It means lean and serene while separating from the fiend.

Out of my desert cabin I walk out to see the stars and I say "Thank you Father". For if you can accept these (free) humble circumstances—not eating and loving awesome eternity—you are promised great rewards. The faster gains tremendous favor and power in public. People are magnetically attracted without knowing why. Saints in history worked veritable miracles while fasting, the lost art. But I tell you truly the fruit and fat makes it simple— you'll be rid of the pimple and even have a dimple.

✂ THE KING IS INACCESSIBLE ✂

To gain the power of which I speak you must become *inaccessible*. Go invisible and silent, for the more humble the more power. Be like the Eastern mystic who won't be bothered with people, arguments or worry when fasting. Let your eyes go off to the horizon—see the mountains and eternity? Transcend silly

superficial society, sick systems and sin cycles. See the light. Become your might when free of culture, clans or cults keeping you down. When finally clear of clods you will see your *entire* destiny. This is your greatest joy and remuneration for it's God's creation. Now you've laid the foundation and despite much frustration you should just feel elation. Use the fasting ideas as your application—you'll see a gene-mutation. All systems and centers open up in supreme dilation. You've been through such desolation in what was a soul-dehydration. Giving into temptation makes life such complication. So now without hesitation, eat right and fast—as all energy pulls upward now you'll see your aristocratic elongation. It'll be your vocation being on this magnificent vacation.

I pray you find your True Talents today as you fast and pray. But I must say: the faster is usually alone yet has many friends. Free of food and the frantic phoniness of society he is the True Self, a symbol of eternity: recognizable to all though unknown. That rarely-seen symbol brings mass attractions yet even these distract from the inner call. The faster is inaccessible and that is his Greatest Power. Avoid the "spirit of familiarity" for even your fans can turn suddenly. Stay free and stay fine for you are rare—the few. That shouldn't make you blue but be your cue for over their heads you flew and that's why they pooh-poohed you. But now you're true at your highest brew—you're a champ through-and-through.

NATURAL CREATIVE SCIENTIST

Most react to the fastarian path with horror. Whatever made us this way? Pressure, the need to find higher pleasure—the spiritual treasure evoked while alone in leisure, not society's false measures--and the inability to live in our bodies puffy, distended and dulled from the creative. We finally learn that nothing ever tastes so good as it feels to be thin. Puffy we cannot stand for we are the type that must fly or die. Get thin, it's "in"--the inner journey all our own. No one can touch us there for it's rare living on air and just watching them stare is no compare. The fast is the best way to prepare for a full mind-body repair, so no more despair: you won't look so square without a prayer. They say we live only foursquare years so beware and take good care. Fast for that castle-in-the-air: being a millionaire in a yacht chair.

◈ THE UNIQUE WOMAN ◈
Pegged and Dropped

The fastarian potentiate is very different from other people and it hurts to be separate. Recall this slogan when down: Take pride, you're on top. Fast: now who's the wet mop? Any multi-racial female is timid to cross anyone's "line" so learns to maneuver around man's legalistic mores--for once labeled, one is invalidated: "pegged". There are always sophomoric family members with psych backgrounds establishing hierarchy through labels, after which nothing one says matters. The major culprits of this subtle crime are laymen but when they collude with professionals the victim may be drugged, institutionalized or stigmatized for life. Refuse all labels and never let anyone analyze you. It's a matter of life or death—reject "those in the know" with this slogan: "I'm the cat's meow. The labeler's a dumb cow."

◈ EMPTY PROMISES ◈

Try to always see the differences between someone's promises and their inner emptiness. The more they promise the more they pretend. Lovers of fast men: this will put you on the mend. Especially the daughters of alcoholics are ever-so-ready to follow this trend. Trusting too early is a sign of immaturity—but oh how easily they befriend! And the males like radar will find this type, and every single time the female--unable to defend--will descend into his unholy den. If this is your tend you must STOP and fast (on men) into your own inner journey, or doom will impend. This was me, my friend—and every single time it was a bad blend. When you can finally see this as a wicked portend (from a love-starved child of misfortune so ready to lend) you will crave so much it's end you'll do anything: Fast and you'll be so strong this sad pattern will suspend.

Fast and pray today: find your own destiny and avoid all "their" projections for each herd has it's own threshold limits beyond which something is labeled as "bad" or "mad"—verbal abuse saddening the glad. I've crossed that line many times and have learned to just stay silent and cordial—fast and pray in my pad but turning away from the cad (in refusing to insult back, I'm a grad). It this latter day era things have turned so bad it appears targeting the meek is a fad but don't get bitter: Fasting is a time of "silent sensuality" to regain that tender heart. So many times life seems hard to take but with the fruit, fat and fasting regime you'll get well as you detect the fake, to whom being properly nourished and fasting makes you impervious—got my take? No matter what the imperious says you're as serene as a lake for you're rid of that snake and the ache he makes—so today, make your break. It won't be hard to take if you avoid the cake and fast (even after steak). You'll be amazed at your elevation when you awake as now you're strong (no more a flake on the edge of a break). You will find this break is followed with a wonderful transformation like a quake: the earth opens up to reveal heaven, rid of the phony leaven. It's been so uneven, these false weights from trust unproven, never being forgiven. Well now you're with God and angels and finally thrivin'.

❧ FAST TO YOUR HIGHEST CALLING ❧
Eliminate ALL: Walk tall, Hear the Call

The people- and food-fast takes us into the highest calling: the treasure hard (impossible) to attain through any other method. When obstructed there is a heaviness to the personality. People are cluttered with the nonessential and useless distractions: like useless or dangerous people we think we need. See that nice lady? She's really a rattlesnake (don't eat her cake). Learn to see through images for there are bombs in those

JUST SKIP DINNER

pretty packages. The champion succeeds by eliminating all non-essentials in mind, body and group. Leave all this heaviness behind: press on to what lies ahead! Rid of people problems you become a sage or a monk finally able to accomplish something. Rid of bad habits you become a prosperous saint and rid of debris (the superfluous flesh from wrong eating) you're a savant with a miraculous body well into old age. Living this way—truly free—will make you ecstatically happy and productive the rest of your long life. Be like a Godfather: totally inaccessible—hard to find, you're a real find. Your final inaccessibility will create mass attractions to the *true* self, not the scared elf. Never call out again—make them wait in line to shake your hand. Get out of environments where you're merely tolerated and stay in those in which you're celebrated. End this old tend and when you see old patterns re-emerging, instantly go back into your fast—that's your divine cast (God's send).

As you leave the chaotic outer world behind and implode into the orderly inner you will become far more productive and useful the *older* you get. In your eighties you will just begin to maximize your mental powers. As the physical wanes the mystical and mental will increase to the same degree—so look forward with glee. All the morbid cultural fears about age are the opposite to the truth. Never forget this expression: OP-TRUTH, for all popular notions are false. There is nothing more pitiable than an elder feeling vulnerable and less-than, out of fear of loneliness: you must always realize you're at the top of your game, as age increases adroitness and even fame.

If this lass was an ass I still have class for if I fast today it'll all go away. So they want me to conform—to what, the norm? Let me tell you the truth just like Ruth: Do your own thing. The pain is so great I know it of late. It's like a police state: they get so irate as they berate--but this is old freight. I had to drop weight for these old pains to abate, and the fast made me straight. It's like getting a clean slate and stepping up to the plate (or finding a mate). At any rate I took the bait by refusing the fruit crate or even one date. It was my fate to make all food wait thus ending the debate. Now I can relate: the fast is my soul-mate, the Grand Estate. I was filled with hate but now I just skate over the pain as I go off to fame. Just tame the flesh and this ends all blame so now you'll get acclaim. You'll have a new flame as to the world it's the fast you'll proclaim.

✥ A NEW RACE: JUST LOVE YOUR HOME ✥

Love your chosen family and live happily. Mine is a dog pack—these guys are my home and we're just like Rome. I was scared for years having been eclipsed by a hostile clan, assuming the reality of other people feeling higher than. Why so long to achieve my own? I first had to learn to love my own home all alone. Here each dawn is a new beginning. It's like a new inning even if sinning. Divine forgiveness covers all—isn't that great? It's like being born with a silver plate: just repent to return to home plate. Here at home there's no absentee-rate: it's just innate as all sorrows sedate, no irritations to grate, no strangers to translate, friends at the gate and all day to create, constant events to update for in our own stream *we're* the head of state. Is there any better way to relate? Your home you must never desecrate for it's your main trait as your talents inflate as every single day your moods elate. All this joy, just because we never ate. The home allows us consistent confidence—that is being carefree: never careless or caustic but ceaselessly creative. Is there a cost? Yes but you'll be the best—that means the boss.

Never let them get you down, for you aren't their clown. You're worried over town? Not a sound: go silent, that's the mint. Enjoy your own home, get my hint? It should sparkle like chrome as divine as a dome where you can meditate (*ohm*). Here you are free to roam, be an elf or a gnome. What great joy—Shalom (I'm talking peace, prosperity, fullness, wholeness, quietness). I felt scorned for two decades, so what were the wreck-aids? Fast-fat-fruit and friends (pets). Pet and Music Therapy are the highest so cats and dogs are our nets—stay close to the furries for they are our loving formulaes.

JUST SKIP DINNER

Now just hold on--for your victory is soon and sure. You reach a point in completion where destiny takes over and even *you* can't stop your success. The foundation has been laid and now God takes over to complete His work *through* you. You've sown the seeds and now harvest time is here. Take your eyes off the past pain and persecutors—they can't hurt you now but your fear they *can* trips you up. Just remember regarding your war: it's God's battle so you won't even have to fight. The world's like a snakebite—without spirit it's a blight. But with the fast it becomes bright--a blissful flight. Yes I was contrite after my fright but now I'm in constant delight.

If you go away (eat, cheat, feel defeat) what would I say? But if you stay and fast we'll find a new way and change what they say. Ignore the many empty face for we're a brand new race. Fruit, fat and fast and you'll last. We're all at times an ass-- does that have class you may ask? The answer is yes nevertheless and all other attitudes I take to task, for since I dropped the mask I no more come last. That's because I have Him—the Big Man in the sky. He's my only tie and that's no lie. I no more cry having this ally: no more to defy or be dominated by, for I have a constant supply and someone on whom to rely--every morning I say "Hi." Then I see I won't die, the day won't be dry and I can handle those gals and that guy. There's nothing to buy just all-day to fly staying so high in the sky. This relationship and the fast is so much better than conversations so dry let alone pie or rye. So I say give it a try and you'll say "aye" or "my!" Don't be shy as all food you deny to look destiny in the eye. It's seeing life through a rose dye--through the fast you're a master spy while you get so spry.

Food—Yuk! It's a truck to have in the gut running amuck like a mutt with a big butt—it's worse than smut! Avoid that rut or just eat a nut—and be so glad again, a thin lass with pure class.

91

7
ARIAL NARROW

✥ NOW IS THE TIME FOR ALL GOOD MEN... ✥

Do you feel misjudged, maligned or identity-maimed? You're in a sick system keeping you down. The superior man *must* rise up, for he cannot be controlled by people and God-Almighty too. The time has come to lose your numb and don't be so dumb for you're a plum being thrown a crumb. Just think what you can become, yet you're being treated like scum! You must not succumb-- nor keep mum--while being seen as a bum but watch the rum and you won't end in slum. Just look where you've *come* from! You've grown much more than some so just continue to hum along in your own stream, beating your drum. Hey--you've got me for a chum and I know where you're at: stuck in black gum when here you have a green thumb or the like in whatever *you* like. But um, do your own thing--the future holds a grand sum (yum) if only you stay away *from*. Hold on, read these thoughts and then meditate: ohm.

✥ ARIAL NARROW ✥

I was always "arial narrow", that is: ethereal from being thin. But still there was a heaviness in my spirit filled with obstructions

from people, habit and wrong eating. Having overcome these useless burdens I'm much lighter now. I no longer want the low-octane world of superfluity, excess and faithless relations—I want to feel the nuclear power of spirit in great depth. I no longer want the herd's dense reality of people, places and things but instead to fly high in this spirit—this is freedom. I learned this through daily fasting when I became filled with creative ideas, energy and vitality: the *fastarian charisma.*

◈ LIFE CYCLE ◈

I'm talking about a major transformation in your life in which you'll soon go from dense to clear. Living in spirit you will receive many creative revelations in dreams and visions—"putting things together." You'll receive continuous inspirations from God. One's life cycle is defined as "the series of changes in form and function through which an organism passes." There are phases to a life and you're about to make a giant transition as you catapult to a new dimension being led by your Father. You can feel it as a faster.

In the spirit you have life force: *elan vital.* You become "life-full": full of (and exuding) vitality. You're about to fall into your force of destiny--being *plum*: something excellent or superior. You're plum color will average dark reddish purple from eating right and fasting. You'll be *plumb*: thin and quick— ethereal (aerial narrow). It's the shortest line, the path of least resistance: the quick to the core, the winning point. "*Plumb*" refers to a lead weight attached to a line and indicating a vertical direction—that's you straight up to God. Verticality is true- -it means you are straight: directly, exactly, immediately

93

completely, absolutely. You'll become a *plumb*: tending to examine minutely and critically adjusting and testing. This is the perfect spirit within, felt in clarity. The *plumb line* is exact, downright and complete.

✌ PLUMATES ARE FIRST-RATE ✌

After being so ugly with worry and anxiety, all has changed, for the ethereal spiritual straight-to-the-point is also *plumbago*: like a tropical plant with showy flowers, a cable that "frees clogged pipes"—you get right to the heart of the matter and clean it all out. The plumb also has *plume*: large conspicuous or showy feathers like a bird. Archetypal leaders showed *"plumage"*--distinctive feathers like a tuft of hair worn as an ornament and a token of honor or prowess: the high prize. It's also a *plumate* animal structure—like a full bushy tail. You're a *plume*: decked with flowers you *preen*, indulging yourself with *pride*. Congratulations!

No more weakness from sin: Being *plumate* (covered with ornamental plumes while being forthright) your boldness-from-purity gets you out of a lot of trouble. When we come up against the enemy they drop or sink when seeing us—this is sudden and heavy: the definition of *"plump."* It also means to "favor someone or something strongly" and this happens as the public and authorities give us favor with "heavy" support and favorable publicity. The plump goes straight down and straight ahead--this occurs "flatly and

JUST SKIP DINNER

unqualifiedly." The "plump" also means "group or flock"—our usual obstruction the transcendence over which we achieve full-blown world success in our field. The encumbrance of people is that obstructive but it's no more destructive because the *power of purity* is more seductive.

✍ FOOD AND PEOPLE OBSTRUCTIONS ✍

We were held back by a *plumper*: an object carried in the mouth—mainly food—to fill out the cheeks. Looking like gophers we could not get ahead. It was like lead what we were fed. We were mis-led to seek out Club Med when all we needed was to fast and go to bed. Then while the enemy gives 100% we give 150% and win hands down—at last, no more dread as God throws us a spread. This all comes after people-problems we shed, as from all verbal abuse we sped. For it made us white—not red—as it filled us with dread. These merchants of misery were not God-led, it was the devil to whom they were wed.

When rid of the "plumper" we become a *plumper*: a vote for just one candidate—ourselves—when two or more could hold the same office. Yes none can compare for now we are that rare. At first we may scare but we have that flair--living on air as we shun the éclair. That's how much we care to be in the boss-chair: we live on plain-fare when they wouldn't *dare*. How much can you bare being as big as a bear?

You're God's heir and together you're a Grand Pair. Eat a slice of cheese and a pear than fast with a prayer—that's not square it's a heaven-powered affair. But beware as they glare though there's no compare (they're in need of repair). They can throw a scare so stay solitaire to be a millionaire. The *plumate* is *pluperfect*: more than perfect, he is past-perfect. The past has been made perfect along with the present and future. That's living in the spirit world—it is timeless, ageless, eternal and wonderful. For "karma" is a golden rope spread through eternity: when the

95

present is soiled it wrecks the ends. When we repent it all repairs: no matter what bad thing we said it's now seen as good by our friends and non-friends.

✒ NOTABLES ✒

The spirit world is *plush*: notably luxurious as we go *plushly* in grand splendor and magnificence. This is *plussage*: amount over and above (we get a *plus* sign). We become part of a *plutocracy*: a wealthy and controlling class of rich men—a *plutocrat*. Evoked is the archetype of the God of wealth, *Plutus*. We become *pluvial*: characterized by abundant life-giving rain. As sudden and massive as a geological shift is this life change: the inevitable changing of the seasons in a human life after being deranged.

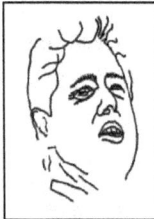

Living on air (invisible food) we become filled with *pneuma* (the soul or spirit) and become *pneumatic*: moved by air pressure--this is spiritual. Having *pneumaticity*--a condition marked by air (now that we're clear)--the empty cavities form a vacuum of mass attractions. *Pneumatology*--the study of spiritual beings or phenomena—describes the basis for these attractions: To get all, become empty! To get to infinity be a big "zero". The *prophet* is vacuous—0 plus 0 equals everything in the universe. Yet all we need for full sustenance and power is a pocket into which some cheese and a few raisins fit. Now we can pay for everything out of our pocket.

✒ THE SPIRIT (ENERGY) WITHIN ✒

The spirit comes through in dreams, visions, purpose and destiny. I can move in this spirit through fasting after being properly nourished with fruit and fat. The fruit cleans as it raises vibration while the fat acts as an electrical conduit for spirit—this is magnetism. See for yourself: begin a Vision Quest--Fast. It's up to you how long it should be for *fasting consciousness* is the point. Let no one pull you down into legalistic dogmas over fasting—the daily fast as a routine is dynamite. Just start and let the creative spirit flow: now you'll be in the dough for you're in the know. Fasting is a time to go beyond the earth to the whole galaxy because you're like Thoreau. Only by fasting can you transcend petty problems to see the spirituality of any situation and what to be grateful for. When the gut-brain puts you in dismay just stay in the galactic fasting state soaking up the ray. No delay—let's see what you say after this incredible day! Eat one high-fat meal and you'll feel tremendous release as obstruction evacuates and then glucagons suppresses your appetite for the whole fasting day. You'll see a huge difference in power-in-public. Get the glow, show for dough.

∽ SAINTS EAT ONCE DAILY ∽

The Saints eat once only--then one day of fasting is like a thousand years to the Lord. Fast forever— by every day eating once and then saying "nothing else, now I'll just enjoy the weather." Say "no" to all: "I'm fasting today." This is the only way to rid yourself once and for all of your biggest obstruction: food and people habits. Then your work will be marked by *lithe*: effortless grace and your features will resemble *lithography*. Now you'll have full life—glowing, burning, ignitable or explosive, energized. You get a means of support and subsistence, a *livelihood*. You'll become *lively*: full of life and vigorous, full of activity, spirit or excitement—intense, keen cheerful and resilient (bouncing readily back upon impact). This is living in a vigorous, energetic or *spirited* manner. When your life suddenly changes to a new form you'll say "Lo and behold" and other expressions of great surprise. Now you may look forward to your *livery*: the uniform of servants readying your horse and carriage, as to destiny you plan your marriage.

∽ VISION, PURPOSE, DESTINY AND DREAMS ∽

Never call out—just fast and *attract in* all that you need from the world. The world says "I'm so great I'm going to make you wait." They're third-rate always coming late. It's all from what they ate

97

☐

JUST SKIP DINNER

as they always take the bait. So they miss your date and that's your fate keeping them around (just dead weight). So end this old debate for they only sedate you, the Great. Get a clean slate and end their berate for destiny does await with success at the gate. You were eating a crate: it made you nervous, it made you hate. That's not a good trait so now let's go straight, all pains and delay to abate. It's like getting the highest interest rate.

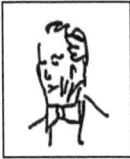

The world is the one with the ego while the faster is ego-free--so he can flee the world quickly through the fast. It's the new life into which you've been cast as the fast takes you to a higher caste, a future so vast while the ugly past, that is passed: always coming last, ever-harassed, controlled by idiots at whom we were aghast. But now what a contrast: Our old hang-ups have been gassed so now new monies are amassed. Now just avoid all old systems or you'll be miscast--them you'll certainly outlast for you're unsurpassed, for at long last you're doing the FAST.

⧬ HAVE A BLAST AND DON'T GET CAUGHT UP ⧬

Have a blast for at last you've found the key. The fast is the way to be, just wait and see! It's like getting a huge fee when free of people- and food debris—you can't possibly foresee your success to such a high degree. It'll fill you with glee, content at your Creator's knee: that's the way to be (it's the true key). It's like a spending spree or giant potpourri all for me and thee—and it's free. The past sins are a dead sea while the future's a jubilee, I guarantee. Fast and it's God's Decree—whoopee! No longer you'll be a would-be or wanna-be, so try it you won't disagree. As you began a trainee, now you're the Marquis.

If you're to have visions, purpose and destiny you must dismiss—not get caught up in—the ways of the world. Live a spiritual life and you'll be able to love yourself: No one really wins with the world's ways as there's always a put-down or more competition to deal with. Live spiritually and your health and

98

wealth will improve remarkably. Do you want to be stuck in worldliness or get beyond gravity? This is spirit, or what Einstein called *energy*. Just focus on where you want to go, not where you've been. For "where there is no vision the people perish." Stop what you're doing, fast and ponder. In complete quiet the spirit comes through to enlighten your world.

SPIRIT GUIDE

Open to the Holy Spirit and He will guide you through all this plus everything to come, announcing in *great detail* what you should expect. Living in spirit means you're prepared for all things through this inner guide—but this Powerful Presence dissolves through worldly involvement and other useless distractions and needless complexities. Think of all the world's petty past-times, while *you* have visions, purpose and dreams. So now decide: no more human control! Have a higher perspective. Act different because you're now nuclear power with a higher guide. For out of the treasure of his heart a good man creates a masterpiece. So what are you waiting for?

I'll tell you the score and cut to the core: their distraction is more than a snore for it makes you sore—but when free you just soar! Resentments you store by joining their corps so what's it all for? Their pastimes are a bore so close that old door. That corny heretofore was a terrible chore but now you're almost to shore and so much more--the blessings will pour when starch leaves the pores. Now clean out your drawers and get ready for visitation: there'll be no more wars.

CLEAN THE ACT: CUT THEM LOOSE, ENJOY THE RIDE

Fast and pray today. The world is petty competition and jealousy while the spirit is total order of magic coincidences. So face the past and repent then press on to what lies ahead. Give up trying to please people or they'll ruin your life instead. Were they ever

there when you needed them? No, so get your priorities right: no matter what you did in the past, fast and automatically get the gold stamp of approval from God, the only One who matters, the Source of all blessings (once rid of people's messings). Repent of all people-pleasings then pick your peers with pride. Just ride the tide and enjoy life in your stride. I'll be your brief guide never to chide, so now you'd better hide from them who pried as you cried. You tried, so now have pride so they can't side against you (being sin-tied). Enjoy your fat meal (even fried) and then fast so your path will be wide. Without the fast into new life you may as well have died.

❧ ALWAYS TRUST YOUR FASTING INSTINCTS ❧

I'm grateful that two days before a friend's stroke I wrote a note of great appreciation. Déjà vu? Yes, fasting is true. This is something you always knew and later you can say "it's the only way I grew." What you've been through! But now to hell you've bid adieu and your reward's overdue— it's in bright view. For now make do, for your life still needs tending to: For if success seems out-of-view then there's still lessons to accrue, mysteries to construe or defects you've become deaf to. Now do you still want to pursue? The sky is bright blue and you've a wonderful hue—it's tantamount to meteoric success you'll come into! With all things made new, just delay your frantic strivings for success and instead just fast, while your own inner life, dig *into*.

Fasting is riding with intuition. You must always trust your fasting instincts even though they "make no sense." Avoid reason: just trust and act now and understand later. Tired? Constricted? Worn-out? Depressed? Angry? Take the day to fast and do nothing else. Just lay down and let the PHF (positive healing forces) perfect the engine. Don't talk, read or worry but just rest the entire day and don't eat. Your reward will show in the morning--as all problems resolve you'll be addicted to daily fasting for life (despite all warnings). For it cuts through problems like a knife and ends all strife so have it be your goal to release sin cycles and sick systems for life! Seeing the limitations of these old ruts brings our regeneration—and visa-versa our regeneration brings recognition (denial lifted) of these old blocks. It's a matter of the clock—as it tick-tocks the gold ring comes around again, and you'd best be out of sin or join the trash bin. In God's universe all is timing for what you've been priming, so fast to keep climbing.

᪶ ON TOP OF WORLD THEN DEATH ᪶

The famous anorexic singer was "on top of the world" before it all caved in on top of her. Realized genius can fall. Her answer was to fast after fat or fruit—that's all. Without that even her determined spirit couldn't offset her years of self-abuse. Had she thinned this way she'd of won the test: to the end be the best for the rest. She'd have been blessed, never messed nor the victim of jest but regally dressed with crowds so impressed. The most wonderful destiny you will have possessed is to fast filled with zest. So to all the depressed: join the club, wear a crest and make a castle from your nest. Without it you'll be hard-pressed especially with much to digest. Fast and succor nature's breast in the energetic elements of sun, air, stars and breeze—these bring out your best.

᪶ LIKE SITTING BULL A SEER AND SOLDIER ᪶

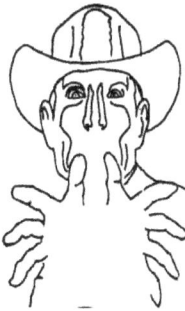

Just be a seer or a soldier. The terrain makes you tough but this training teaches you *true* genius survival. Learn to insulate from outer influences and wear a hardhat—make your mind impenetrable to irascible and down-putting influences. Then make your home like Rome—it is holy ground, your only protection from ignorance. By staying home you see what needs to be done so in time it becomes filled with exciting treasures and abundant adaptations. This is becoming truly rich . Then become the symbol of resistance: *Never* give up your effortless bents without a fight. We're talking impossible odds the genius wins by taming the instincts--being a knight. It is self-mastery which releases this powerful spirit and the result is quality in everything you do. Now showman politics— that brings downfall—but God's glory illuminating and perfecting all work. Your worries are through or only a few, because that starch you deleted was like glue—bringing the flue, making you blue, marked with poor hue and pores filled with goo. Join our fat-fast crew and you'll have complexion like dew and rich thoughts too.

❧ WHERE CARCASS LIE BUZZARDS COWER ❧

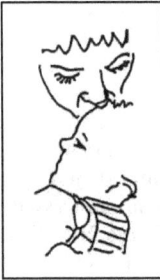

Where carcasses lie, buzzards cower. Watch out for evil helpers for when we get down people treat us as clown. It's the story of Job when his smug friends came around. When they put you down don't wear a frown but bear a silk robe, trust only God and forget the clod: fast, pray and wait for revelation to retrieve success. Never act, wait to be *led* until you see the ghost-like adumbration of your next steps. Soon the superior man will be putting everything into proper place. Always break from your tunnel-vision to take leisure by look to eternity: the elemental reality of sun, moon and stars. This is a higher level of living and good riches are your inheritance. Enjoy nature first then come back freshly endowed with new powers when "every tongue shall be shown to be wrong." Fast and you'll not be put to shame. As you rise up people will always have a problem with it--evil tongues will wag but soon have none effect for God says "they *will* gather together against thee, but not by me. I shall condemn every one who rises up against you." How reassuring--yes this will happen but if you refuse to fear it all dissolves, having passed the test. You'll be riding in high places: the Rolls Royce is made for the children of God so picture yourself riding in one, having jumped over the hurdle that all men must face to win the race. Despite all your loss, this last test makes you boss.

❧ PRE-SUCCESS CRISIS ❧

JUST SKIP DINNER

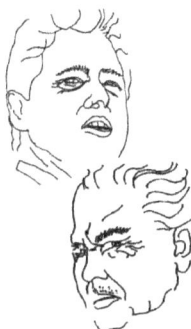

Right before success expect a *pre-success crisis*. This may be a sudden pruning (or chopping) of your life or a relapse into old cycles or systems causing symptoms. The Saints may take five steps back but then ten up--after slipping into habit just recover by eliminating two or three more. Tribulation precedes success which comes from self-abased humility while arrogance precedes downfall. Don't become scapegoats: avoid all boasting, butt kicking and bloat: that's the world's boat you must avoid, for you'll only make it through a higher course-- the universal archetype of all leaders that is immediately-recognizable yet unexplainable dignity. Siding against you will always occur if you stay in sin--the *fallen hero syndrome* is no fun. An humiliating slip can make you a scapegoat for life so reject all systems holding you down as clown. After all, a slip lets you know who your enemies are: Before there was cordiality but strife underneath, but now your relapse brought everything to a head (as your foe made his own bed). From him you were bled—he wanted you dead and living on bread. You were filled with dread all from what you were fed. But fast and that weight you'll have shed. You'll be happy instead, a color of deep red. As your skin is retread you'll begin to get ahead—your future's a golden thread. You're a thoroughbred whose image is soon widespread.

❧ THE FAST IS THE END ❧

Now THE END occurs--the completion of old cycles and the beginning of new ones. Release obstruction and you become energetically childlike every day of your life. The minute your work is complete there's a gravitational pull to materialize it on earth. It took you decades to form the kernel which now attracts pollination, and now you're in a new station. Your work is causation so you'll wear a carnation, as it's all creation without cessation. The past was damnation and deflation—you took their dictation. But fat brings dilation then their donation. For the duration you'll be in elation and then—floatation. You've laid the foundation and it took so much frustration! It was slow this gradation saying "no" to temptation. It's not starvation but it ended stagnation and oh—the *sensation!* Next to God it was your salvation and inflation. No problem your geographical location, the issue's the *tissue's* mutation. Before you were negation on probation, but now it's on to oration. Now for summation: let's raise your vibration then go on vacation.

✒ THE END: THE HIGHEST BLEND ✒

Today my mission ends on the way to destiny. Destiny is discovered and then decided. That makes life really exciting. Get to the buried treasure of visions, gifts and talents so your Inner Man creates magnificent things. Get to your spirit to see what's inside. Did you gravitate to certain things at ten? Find the bull's eye *then*, aim the arrow and let go. You'll see the old life blow away as you stay in the fast today. No more skin of clay being gray (hair like hay). Hey! Fast and pray it'll be just like play, no more their prey. Now I say never stray--it is they who will weigh. Your problems allay they all go away. Instead a new array like a bouquet. But if you're stuck at café or a buffet all men betray. Fast, you'll enjoy a chalet. Stick to the fruit or cheese tray then the fast is child's play. Avoiding cliché, I wish to convey: starch means *decay*. So end all dismay and no more delay. You'll be on display as they read your dossier. To end this essay: After enjoying your filet, fast today. Don't give way by going half-way but obey what I portray, okay? They'll all repay.

P.S. Rest assured that your Completion forms this kernel creating the True End—the blessings of the new life of due recognition. As your self-expression hits the mark it makes an amazing new dent in the world—the

Revitalization of Culture.

Soon you'll be saying: "What difference past mistakes when I can have all of this? They mean nothing—they do not exist. All that exists is this day and the bright moment I'm in."

8
Synchroni-City

Creation is a beautiful design of unbelievable complexity evincing total order. You can hook to this power with total order in your own life by sticking to pure *essentiality*. Eliminate all non-essentials and stick to the *quintessential*--the "pure essence" of who you are. Then you will experience continuous miracles in your life—magic coincidences fitting together for your benefit. We only have to be clear enough to apprehend it all--that means eliminating all distractions.

Through fasting we can really "see" synchronicity--or how everything fits together like a jigsaw puzzle and what needs to be eliminated. Elimination is the basis of all recovery so that now you may attract to bring your world success. It all depends on our level, for synchronicity can be good or bad. If you're at level "1" you'll attract low-minded pettiness and sterility. If your at "10" you'll attract the same to your domain—your own glorious miraculous Synchroni-City. You must cut loose all losers, lechers and liars around you. When in solitude or a good relationship the good make a magic life in one miracle after another. It all depends on the condition of your spirit *and* the spirit in your associations. That's why the bad fits together in extremely sick situations in human systems. If your spirit is stuck in low-minded lusts or vicious view-dominators your world will come together in the world of scapegoating and spirit-killing by other people:

JUST SKIP DINNER

treachery, betrayal and "accidents" shall also fit together perfectly! If you're high you'll be able to create things into being: "calling things that be not as if they were." Let's compare the two states by looking at what happens through conformity vs. solitude. Before the genius is recognized and remunerated he's seen as "strange". That's one of the hurdles overcome for success to occur--one must be sure of himself and that comes from not taking the lure.

❦CHILD OF ALCOHOLIC SYNDROME❦
Love-Starved Compulsions, Obsessions and Antipathies

The love-starved trust too early that any man calling her "sweetheart" she instantly dreams of marrying. Then when he grows cold he's ready to scold, scaring him away for good—"that hood!" For she is still a baby fighting for survival in the hostile environment—the man's lost interest is like murder to the infant. She is a child of misfortune, a mal-adaptant to a vacillating childhood environment--and it refers to any pattern: gambling, over-eating, social addiction, workaholism but is most pronounced in the flip-flopping alcoholic home. Without ascribing blame to parents, we just look at it in terms of *biology*: the only mal-adaptation to vacillation is stunted emotional growth as the intellect progresses--producing a brilliant baby with *bad attractions* from that point on. Nothing can change this pattern unless she develops inwardly. For, being empty, she will attach to any drama "out there". Instead, she must become *rare*, developing her own talents and bents lest she be treated like two cents. Now whole she will attract men to her True Self, her own Synchroni-City--for she expels divine scents.

❦ YOU'LL STAY ON TOP SO NEVER STOP ❦

While coming into his own, genius stays a stranger in a strange land. It hurts to be different or reclusive in this culture. The white man was bent on converting the Indians to their conformity and nothing has changed. But the minute you conform the moment is lost. You've got to stay in the moment to see the miracle--but in refusing to attend to worldly distractions we find ourselves sided

107

JUST SKIP DINNER

against. It's ok if we live on a family farm of familiars but what about the rest of us who must adapt to suspicious misjudgments? All we can do is fast and pray each day. Then God allows us to stay on top no matter what they say, come what may. If you fast in silence you'll be in perfect synchronicity and nothing can disturb God's plan. Just persevere for one thing's clear: soon an inversion of *all* systems will occur. Here the bottom becomes the top and the top the bottom—that end is near my dear so have good cheer. Now lend me an ear: fast for no more fear nor crave of beer. Fast to get in gear as quick as a deer. Do you hear? You can't come here but by the mere fast we are near. Your rear resembling a sphere will soon be sheer in less than a year. Now fast—not a tear, for never again will you be in the rear.

ஆ PAIN FROM PUGNACIOUS POLITICS ஆ
And Petty Prideful Persecutors

It's a *viscerotonic* culture—this is the "social" temperament. You must be social to get along. It's crazy and will make you so. Stop worrying about your reputation with people and start thinking about your rep with God. Remember it can be good or bad, the way things fit together: Bad synchronicity comes from sin of habit or association. The faster must avoid the pure petty parochial politics of his peers and other prideful past-times. Eschew all pettiness lest you feel pain from their pugnacity and lose all synchronicity. It's a pity what comes from the bitty so just hear my ditty: you gotta be gritty yet sweet as a kitty. Rise up: Fast to be witty and oh-so pretty.

ஆ NO INVITES TO LIARS ஆ

108

JUST SKIP DINNER

All the saint-genius-ruler can do is balance factions all day, for there are so many around him. No man is an island: we *all* have to deal with the thugs. Those with the mugs always pulling rugs out from under you (they go so low). But we rulers must stay high, i.e. cordial—but without tie lest we cry or die (we want to fly). The foe's so dry—so bid him nigh and don't be shy (for that guy's *so* sly). Why? He eats rye while we're in the sky. Now *apply*.

Never get too familiar: it's an invite to a liar (these are the ground rules to kingship). The point being: no matter how high you get there's always crap to map and zap from the scene. Then while you nap you snap to your goal while the foe falls in his *own* trap. That's my rap—you gotta cap that ol' chap.

❧ FOOLISH FLATTERY FAILS ❧

Saints avoid all foolish flattery of supposed superiors. The superior man bows to no man, only God. Fast and pray lest they mock and degrade your True Self. This divine blueprint is inferior to no one, for it must be free of social design to illuminate.

You can't apprehend miracles while kissing up--you're no pup! Those caught in the sociocultural mazeway have a brain filled with cobweb-illusions: who said what to whom. Boom: Bomb those lines of interactions out of your head for they're like lead. Transcend to spirit—go light for only that is right. Let the creative alight, then the foe's contrite: a mere fleabite to your great delight. Despite your dull site you're now dynamite, so now just be polite: don't ignite or excite but be *forthright*. Only boldness from purity is right, then you'll end this old fight.

❧ AT YOUR PEAK WITHIN A WEEK ❧

Leave them in the dust to join the fast lane, or bust. Let me tell ya' about lust—debunk it, or rust. Once you're clear you become unique: that is sleek. You reek? With the fast it's all dissolved when you take a leak. As fast as I speak it will narrow your beak: skin like teak no more lookin' like a geek. If you'll seek you're at your peak within a week. Besides the physical benefits to fasting there is *synchronicity*: that beautiful bountiful right-brain reality where it's all magic coincidences coming together for the benefit of the fastarian. Once there you never want to lose it again. This is synchroni-City: boundless miracles in your own home which are lost through people, habits and food.

✍ REPENTANCE ✍

Repentance is an exciting concept: "Pent" is the highest and classiest apartment at the top of the building. "Re-pent" means to *go back up*—to your home-base: being *high*. Repent: turn from your sins and don't keep slipping back only to make resolutions again. That's a no-brain. You slip, they pull your chain—that's a severe drain even a weight-gain. End this pain, for you must now start to reign and it's a whole new plane (you won't look so plain). For you, only purity is sane. Your sins were a stain but once pure again, now the foe's slain. Without that affliction, you're not so vain and you see more to attain. If you can continue to abstain there'll be no more complain. Fast and delight in your *own* terrain—much higher than cocaine: I can promise you there's never a need to call out, just enjoy your own domain! All else is really insane (so very mundane). So at home you'll remain, and fast and it's your head that entertains like a castle in Spain. Make this day your *new* birthday: fast and pray. Fast for the worthy desires of your heart and then prepare to receive them. Behave as though success were a walk-in today. *Once you've become empty the created vacuum attracts it all in.* Fast for synchronicity: make yourself "apt" to receive.

✍ THE SHADOW OF HIS WINGS ✍

What God has blessed Man can't curse for His blessings are far higher than man's cursings— He is higher than what you lose sleep on. Forget man's put-downs and prideful persecutions. Your Father wants to elevate you *above* all that into your own pre-arranged destiny. Take refuge in the highest—in the "shadow of His wings"—until all calamities passeth away (that's the only way). If you are His, no matter what you do he hides you under his wings—isn't that wonderful? He has things all pre-arranged so if you slip up on your own self will--if you get on a silly tangent--He makes it null and void. Your sadness will allay so your talents can convey. But are you *ready* to display-- are you photo-ready yet, can you *dare* compete? Only after releasing all uglifying obstructions are you complete.

Rejected? Scorned? Let fasting turn it all around. Let the ground become the figure and the figure the ground. Then out pops insights for growth and solutions to all life's problems. That means: you become the top and the top becomes the bottom. Fast and let God throw a banquet for you in the presence of your enemies whose faces are pushed in the mud. This is: *enantiodromia*: the inversion of all systems with you on top of Elmer Fudd and his bud.

◈ DESTINY DISCOVERED THEN DECIDED ◈

Raise your synchronicity by your own behavior and repentance: *Elimination* brings success. First you must *discover* what your talents are— those "bents" evident in childhood. Then you must *decide* to pursue those inborn talents with learned techniques. Deciding your destiny means eliminating the obstructions to it's realization: people, habits and debris. The rewards are great: there is no greater joy than making a living on what you do best, for that is effortless. And God will bless so it's marked with finesse: this is True Success. End this mess--your distress all comes from excess. No more regress or people-obsess for it's these distractions that created the stress. But say "Yes" to noblesse and you'll submerge into the elements (just the wind to caress). Now, that's progress: as your mind opens to higher realities never seen before you'll lose interest in "past-times" like chess. Become receptive to these higher inspirations, for the fast is your new dress--all pains to suppress.

111

JUST SKIP DINNER

✍ THE CLOWN: ALWAYS RUNNIN' AROUND ✍

Look at them—they're all so busy and you aren't their priority. Busy as bees without the accomplishment of fleas. They talk of piffle concerned over trifles always lie-filled mired in the trivial. It's a busy life of short attention spans flitting from one activity to the other. There is no synchronicity here and if obvious miracles occur it's "just a coincidence." You have to know you're the best--that's the test. They'll always try to get you down when they're the clown: that's the town, forever running around. As a fruit-fat-fastarian faster you're ever so brown so just wear your crown. Fast for the sense of a hound without frown nor sorrows to drown. Hear that sound? It's your blessings abound. After repentance just ask around—they're all astounded and your foe has backed down. As you expound they're all confound.

FEMALE COMMUNITY
The Biggest Impediment to Female Genius

"Why I don't like people and prefer dogs"
by Karen Kellock
Also Known as: "Why People Make Me Sick".

JUST SKIP DINNER

Because People:

1. are always establishing hierarchy.
2. are always using leveling devices, creating status-tension and aggression.
3. are always catching us in old webs.
4. tend towards idolatry: worshipping sacred cows (money and property) rather than morality
5. bully through exclusion (gossip), a practice most prevalent in the female community—the biggest obstruction to female genius.

❧ GOD CALLS IT A MINE ❧

All that matters is our own reality and personal relationship with the Highest. Adapting to the reality of others brings depression so never let anyone define reality *for* you. I'll give you the clue: it's your own home that makes every minute pertinent. Only this is true so if you can give up seeking outward fixes, you'll never be blue. The only way to true success is *your unique hue.* Am I filled with spite? No because in self-containment I'm filled with might, as bright as a light. I used to get tight and those I knew were tight as a wad--in mal-adapting to them I felt "odd". When they said so did I get low? Oh yes for in having to toe *their* line I lost my shine and the reality which was mine. This higher inspired reality is what God puts *in* us—and other people only dilute, cheapen or crush that. Keep to your own, that's staying refined lest you decline that Great Reward. In the meantime, don't slip to just fill up time—too much wine is not benign. Instead bring out and be the best of your *bloodline*: now that's divine. Fast to align with Heaven, that's the Royal Design. It's a sign you have spine, and your work will be defined.

❧ NAP ON A PILE OF STYLE ❧

Maintaining the True Self is like a war. So let me hear your roar for your *own* lore right to the core then give me some more. Expect

113

JUST SKIP DINNER

it: *tribe-ulation!* That way you'll be ready and never get complacent. Your success is nascent, the new life adjacent. These are the mean latter days where cruelty abounds yet goes unnoticed. Who knows? Your gut, so don't subdue it with a rut. Listen to these instincts and act on them. Let them call you absurd, just fly above like a bird. Your destiny was blurred but haven't you heard? It was all the herd who called you a nerd. If it's just you and God, they all come slow third—*that's His word.* Fast, then you're preferred (your sins never occurred). Don't worry what "they" think and relax for they'll only see the Genius of your situation later. Run that last mile then nap dreaming you're on the Nile on your beautiful tile—your *own* pile of style. You're so versatile, so fast to beguile the hostile—pass this trial and it'll all be worthwhile. Walk that last mile for in just a little while you'll be wearing a smile. The higher your destiny, the more God may want you to be alone in the preparatory stage. So don't feel blue, for no matter what you do, *nothing outward* will be coming to you (save a few).

ᔐ NE'ER MIND THEY'LL ALL BE RIND ᔐ

Having refused to avenge yourself, now *God* takes control. His justice is real. "Those who curse thee I will curse. Those who bless thee I will bless." Run the race. Wear a poker face, hide emotions (not a trace). This stuns better than mace—they'll all now erase. Never see them again your so-called "friends". This will mend your broken spirit, amen—they'll never be able to rear it, again! Ne'er you mind: they'll all be like rind as the wind blows them away (if you fast and pray today). Your fears to allay, no foe can betray (no way, Jose). Now just come inside to True Destiny. Make your cheese soufflé so dry skin's passé. Protected in your castle, the foe's become ash gray—he can't meet halfway while you fast on the Lord's Day. Those who foul play end in sick pay.

ᔐ IT HURTS TO BE IN A CROWD ᔐ

JUST SKIP DINNER

It hurts to be sided against by the crowd. Racism and sexism are *group-determined*. Insofar as one's identity is stuck in the collective he can't see the individual--only stereotypes. It happens more than we know (it's a real blow to see how low groups can go). Don't eat crow, stay in your own flow. Knowing your foe you can just continue to grow. You're sure to glow for you're a pro. They're all soon to know you so fast to show all those below. Then God will bestow just like centuries ago: it's a *quid pro quo* as you'll rise up on your own destiny's plateau, all aglow.

They're all a group due to *sameness*: they've all conformed to some "norm". Since you couldn't and wouldn't you naturally found yourself their target. But fast and all this becomes irrelevant for you've hit the bull's eye whereas before your victim status was a near-hit, the failure's fit. Go solo and heal thy wounds--that's the highest. Only the true champion aspirants can do this, but if done they'll all dissolve like vapor. No need to taper—give them their walking-paper in return for their caper. Start your fast now and wait on the Lord for payback time, good or bad--you've got the world on a string if there are no more strings! Gifts God will bring, *unless* you cling to that fling. You felt guilt and shame *because* you were bound with a nose ring (a recurrent bee-sting). But under His wing you're the Sun King--and it's the real thing, this upswing into spring. The most important thing to remember is: God hates people-worship (idolatry). He's a jealous God because *He and only He is King.*

❦ RELAX—THEY'RE A WET MOP ❦

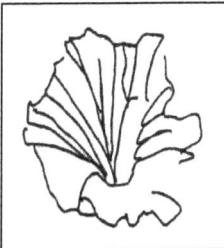

In any group strongholds develop as they *side against the Great*—he who stands alone on his own. He who chooses what gets into his mouth, bloodstream, thoughts and environment. And they hate him for it: putting their low-minded collective corn in jeopardy. So have a party my pretty, it's your own Synchrono-City (let *all* come to you—that's my main ditty). The others you'll pity (the devil's committee) so make your break

JUST SKIP DINNER

and don't look back. You've overcome your Self so now take the land. Let them band together: they're a wet mop but you'll stay on top. How? Well what can I say. Just fast and pray today. Enjoy thy ray and no more dismay over all their betray, the delay or your felt decay. Just finish your work (carefully now—like a crate of eggs) and be patient as the seed brings harvest and God gives you a bouquet: a veritable spread like a buffet. Never be alarmed—your life will be charmed. Fasting, you're well-armed. To Hell they've been farmed for all whom they harmed.

ACCEPT THE RAY, THE SPIRIT NO STRAY

Just say "Lord help me in this situation: give me vision of what is happening and everything that is *going* to happen." And He will. When you again see the swill the Lord will've put them through the mill. What can I say? Just fast and pray today as all goes away. Just accept the ray then the spirit no stray. Our joint fasting is a prayer center--the herd won't enter where God's the mentor. If we are repenters they'll be mere renters with us the land-lenders. Now no more benders attracting pretenders—those awful officious offenders! Fast for angelic defenders: Now we're contenders as the foe surrenders. No good and bad blenders—when free of sin-vendors, now we're big spenders.

PROBLEM FLIES LIKE THE ROBIN

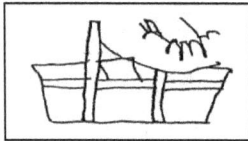

When you get a "no" answer in prayer don't dismay. Think today: come what may if I fast and pray we'll get a thoroughbred from a stray. Tho' feet are of clay the foe's kept at bay—he's getting' so grey. This is my essay: If you fast and pray you can even just forget work, and play: It'll all take care of itself, so don't go astray: Don't fast to low-weigh, but so the foe God will slay while making your work all-ok. Fasting you can expect with certitude that all-bad turns to all-good in your favor with you on top. Can you hold back a tidal wave? No, so now you'll be the new world rave. Though we live in a cave we're here to save the multitudes from the vicissitudes of food and the demons' squander and plunder. Lay down your head and meditate so you may levitate—above this problem which flies away as a robin.

⇜ BE BOLD, BE BLUNT BE BRASH ⇜

 Be bold, be blunt, be brash. Tell them what to think but never listen to *them*. Be *undemocratic*. Choose upward--you must! Wipe off the dust, or rust. Then you'll be saying: "I used to be sad but now I'm glad. I used to be mad but now I'm *bad!* This isn't a fad--it's your launch-pad. Through all you've been told you've been had, but now you're a grad at home in your pad with God your Dad. And I might add: the hangers-on He forbad. You got it my lad? Now just create on your scratch-pad then place a news ad.

⇜ TIMING IS ALL: PEAK PERIODS OF DAY ⇜

 Everyone has a peak period of the day when they're most creative, and when they should eat. My best time to work when creative ideas flow is 3 a.m. to ten and I eat before dawn. We all have those peaks—finding your perfect time can transform your life forever. We're talking time: our own peaks sublime. Get clear of other people's time schedules for it's a crime to miss God's design. Coming into your own destiny is like cleaning away the slime, that awful grime. And here it's totally from the fast (for less than a dime). Now together we can climb, forever at our prime.

"Attractive, clever, appealing, bright and kind—yet a tragedy." Here genius ends and madness begins. Repent of sin and these "crazy" symptoms leave. Your skills take you places but your fasting-built character keeps you there. Believe God and accept all He's prepared for you, then you'll be healed, free of that plague. Go from intelligent and dull to bright and explosive. Repentance is like turning a light on where there was utter darkness.

9
GENETIC CLAN

Talents or Troubling Traits?

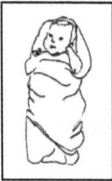

Having cleared the blood body and behavior of all superfluity you take on what is called *genetic body fluids*: the *best* of all the traits, genius and talents in your bloodline--from the beginning to now, it's all accumulated in *you*. If not this clear you may take on the opposite tendencies: *dysgenics*--the *worst* of all these traits. Just like synchronicity it all depends on the state of your mind, body and associations: If clear we attract good in our genes and circumstances, and the opposite holds true. Look through your bloodline—what traits and talents can you relate to in your favorite relatives which are also in you? These are present in your seed symbol but need prodding to come out. But also which traits do you see in your least favorite relatives which are also in you (when down, a clown)?

A genius is usually different from his family. The other members may collude together chiding his differences so his mistakes are magnified (it's a losing game). It may be only separation/actualization and fasting which can spring him up to the top. He's a wounded soldier—it's a damaged crown seen as damaged goods: "there's no hope for him". But the good news is that's not true.

JUST SKIP DINNER

THE MOST POWERFUL FORCE: DESTINY

The most powerful force on earth is your heavenly destiny—God's plans for you and you alone. Unknowingly the world will make way for heaven's plans for your life, for creation is built around *you!* Whatever God made he did it for what he wants to accomplish *through* you, and *at just the right time it links with history.* Your evolution is a byproduct of creation as everything created fulfills a predetermined function by God and will also help you *fulfill* that plan. Put on all else a ban. When it comes to letting self-defeat go, do you even think you can? We're talking a whole lifespan or being as simple as you began (that is being a wise-man). We're talking holy—the True Self is better than the fake (the Peter Pan). Modern paleo science shows fruit, fat and fasting is best for man—so it's the way to win health, svelte and *then* the fans. But first we must elevate above the family tree—the clan:

BE ENTERTAINED BY YOUR MIND

No matter how "high" the source of entertainment it doesn't compare with True Reality—that's your destiny in each divinely-planned moment. You can easily miss this mark of the highest calling. Addictions (TV, food, drugs, people) *magnify* because they do not *satisfy*. Energy misapplied to sin cycles or sick systems make you a loser not a space-cruiser. So fast: open your mind and just wait for instructions.

You'll *snap* to your destiny with no more delay, for there's an anointing to obey God now—but putting it off will dissolve that power. Procrastination means not obeying destiny-instructions *now*. If you refuse or rebel you'll be "devoured by the sword...because you delay or turn a deaf ear." Don't put things off—that's a lazy, slothful, lethargic spirit. When God speaks, obey. For if it's your destiny to be rich and famous (in order to

119

make a new dent) all else brings misery. Listen to the inner voice: if He says to take the trash, pay someone back, give something away—obey with nothing else to say. The reward will snap in and it'll be a great day (a grand display). No matter how seemingly trivial—obey that Inner Voice for before a product is sold to the public it must be tested. You must or rust: obey for success or bust. For now is the time for you to come out of obscurity to the fore--get ready to take the floor! Your whole life has been leading up to this door, for we're in a war against low-minded debris we *all* should abhor. Why long for things that came before (fantasies galore)? Just go inside to gain rapport with Self and God. You'll soon be saying "the world and flesh no more." Heretofore you were stuck and down: (you wouldn't want to *see* all that came before). What is more your failure was a template hooked to cycles and systems—this I must underscore. Here's a fact to brace yourself for: Only through the fast (on people, sin and food) can you finally know the score.

✎ ALWAYS USE YOUR PAIN AS INSPIRATION ✎

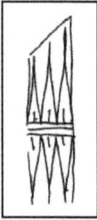

Don't be a pathetic has-been squandering your talent. God gives talents to those using what they *have* and from those who don't he takes them away. So no matter what, work: always use your pain as inspiration. *The Greater one is, the more frustrating his path.* As the Judge said to the great Patrick Henry the greatest orator in American history: "your gift of speaking makes you dangerous." How can one win? The herd sides against the great until his time has come (when he's first-rate)--then for *them* it's too late. If you stop your work by asking "what group will this appeal to?" you're sunk. You'll never make it by considering your audience first. Just fast and act—let the spirit come through: this is Godly tact. Elevate them so they can appreciate *you*. But don't get too close--for they're disloyal too.

✎ REJECTION: SHAKE OFF THE SERPENT ✎

Evil uses the pain of rejection to keep people from going forward. Shake off this serpent! What to do about snakebites: Don't let their poison ruin your life for with every new opportunity there is opposition.

JUST SKIP DINNER

Peter told Jesus he would never deny him--to which the Master replied: "[don't give me that.] by the time the cock crows you'll have denied me three times"--and he did. Don't put your trust on temporal things or passing beings. To stay ecstatically happy, eye only the *eternal*. For in the case of all Greats you must learn to stand alone (even if you don't need to) for Your Day Will Come— it's etched in stone if you refused to be a clone. Share in His suffering then share in His glory--it's the same old story.

So now divide: from flighty, fickle, fair-weather friends and petty partial party-time peer-induced pleasures and place your whole hope on God and Eternity-- these are pleasures forevermore. Pay the price, then stand firm. Bite the bullet: fast and pray today come what may. Now without delay go beyond all you dismay: the meanness of the trivial in everyday life. These pains will all allay as your troubles fade away the minute you obey the call to fast and pray. In dense energy your life's been in disarray— it was a terrible decay as you were the prideful people's prey. It wasn't from what you weigh or even how you display but rather looking the other way as they *always* did betray. Just relax, fast and hit the hay for soon it's judgment day. The fastarian life glides like a sleigh as brilliant as a Monet.

✍ PALTRY PEOPLE PROBLEMS: PERSECUTION ✍

We mustn't let people devastate us, for a crowd is crazy. Out of envy and malice they group together like one mind against one, agreeing—*consensually validating*— the victim's inferiority or evil. Become whole: you will win out as higher always gains victory over lower. The word "rejection" means to "despise, set aside as having no value. To loathe, spit out, cast away, give no place to." There is no greater pain until you see the true gain: for the cursed genius that's the main--let me make that plain. It helps to know they aren't rejecting *us* but rather God

121

JUST SKIP DINNER

who sent us. Once God places a call on your life there is no greater satisfaction as you become rare and different. At first the herd hates and slights God's mouthpiece but you must never be upset over the serpent's bite again—shake it off to be King. If you are bitten and don't shake it off the poison penetrates your system and you become as bad as the evil that sent it.

✥ SWOOP NOT DROOP ✥

A magnified template makes one turgid, slow and "heavy" as the eyelids and ankles droop more through the years. This is ordinary (dense) aging in contrast is Oriental Aging: getting lighter, purer and wiser with age. This comes from continual elimination in order to get more of Eternity (Einstein's main interest). When you fast, instead of the "big droop" all lines "swoop" up as the elder sees his source. From all else you must divorce then you're free of remorse. What I mean is a total change of course: make fasting (on all obstruction) your vital driving force for only after overcoming herd-obstacles can you relieve yourself of people-burdens. Since these are often family members, transcending the block reveals the Genetic Clan—the bloodline's cumulative talents in you alone: Now free of family-failure he goes off to fame and then the clan wants him back (that's always the game).

We form "templates" when young—these are the system-scripts to which we must adapt. The brain makes a *neuronal model* of the sick system—*the template of us as one-down*—which it then repeats by rote with others throughout the life of neurosis. These templates must be broken as they get progressively worse with age. At some midlife point we either break out and shoot up to our eternal destiny or the templates become magnified and heavy as we become *ordinary* (age and die). In the latter case the body becomes just hardened memory and degenerates into death. The greatest people in history were shameful failures in their early lives having been hypnotized by a sick system holding them down, until they broke out and shot up to world fame and greatness. Get free of templates. Take the opposite view for only this is true.

122

JUST SKIP DINNER

Whatever "they" say, you must think "OP TRUTH" (then no more blue).

৶ THE INDEPENDENT MIND IS REJECTED ৶

"George Bernard Shaw once said "now, only as a writer have I found myself." This author can relate to that completely—without my work there was no grounding in life as I just attached to passing dramas turned treacherous (the mean and the lecherous). It was a tragedy like leprosy, until writing made me human by hooking me to God and eternity. When that happened to Edgar Allen Poe he said to his rejecting father: "The world will now hear of the son you thought unworthy of your notice." The independent mind is always in opposition to society, man's traditions and religion. For the independent mind the issue is to be so busy *doing what it likes* it doesn't have time to worry over hostile people. To find your independence and easy life's work you must fast—on people, habits and food. Get ahead for once. Stop spinning you wheels. Over idiots you've been head-over-heels, so now start kicking up your heels and until you're a faster avoid all deals. It's just a matter of deleting meals to increase your next-day's appeal. As the past your fast conceals now you may advance ideals. Those slimy eels! They're all gone along with all other trying ordeals. So forget the perilous past--it's only the future we want in all the glorious beauty the fast reveals.

It's a common pattern: the rejecting childhoods of writers which is (after overcoming these obstacles) later expressed through a unique, unheard-of style. In this way they are seers (having seen it all) and prophesiers—accessing the future through hard-won wide-angled vision. They've transcended the parochial

123

prideful puffed-up past-times and powers of the petty--which can be very mean compared to the humility of the universal. Writing (and hitting chords) enhances the damaged self-image of one who has overcome these impossible odds.

◆§ STAY FREE, STAY FINE ◆§

Just because someone pursues his talents does not mean he automatically breaks early templates--but he *must* for true lasting success. Most genius falls down unable to transcend the silly childhood squabbles encoded in the neurotic interactional templates. Corrosion sets in when bonded to the past until a certain stage is reached where one must dissolve templates or die. It's not enough to learn techniques—genius must set his mind on the future, forgetting the past totally and he cannot do this by blocking it out but rather dissolving the templates through fasting. To not-eat or people-compete (his usual devices to avoid anxiety) he falls back to *face* and *feel* the past. This hurts but soon it's gone forever. He has a good cry then goes on unencumbered to the fantastic future. The champion must be a totally new person with new reactions to the universe. He can do this by looking ahead—not obsessing with the past (for the dead feels like lead).

It is *early trauma* which stops progress: all "crazy" words or actions are from past templates (like an introjected parent heard as "voices"). Early trauma sears the nervous system and stultifies the psyche at that very point: In fear creativity stops, a template is made of the situation and one acts by rote (based on distorted conceptions) from that point on. The object of fear (mother father husband) is swallowed and reappears later in "autonomisms"—out-of-control words or actions. Like another person inside, this robot circuitry is the opposite to creativity and progress. Dissolve templates or die.

JUST SKIP DINNER

∽ CLEAR OR DENSE AGING ∽

Neurotic man acts by rote (introjected template) and this is the basis of prejudice, aging and death. There are two ways to age: dense or clear. Dense: gravity pulls downward as history (template) is magnified. Clear: gravity pulls upward as templates are broken and now one can "smell the roses." This is coming out of robot-circuitry (a hypnotic trance) to True Reality in the land of abundance. Now you can cut this process short—no therapists to court—by the fast (the cosmic dance). Just fast, transcend (cycles and systems) and it's so much better than chance. Revel in your high-fidelity senses like the culture of France. Age upward—they'll give you far more than a glance. It's your True Beauty that the fast is sure to enhance. It's so exciting (like romance) and with a massive advance (we're talking high finance).

Neurosis is: the past trashing the present. It is acting in the present as if it *is* the past. Owned by a template one is destined to repeat the same old mistakes causing anxiety and avoided through sin. As time goes on memory is stored beneath the neck in holographic "images"-stored-in-tissue. Man is an embodied experiencer—perceiving not the world "out there" (like the clear faster) but rather his own body-bricks from past slights. These aren't the higher-than-kites but the "dark knights".

∽ THE ORIGINAL SYSTEM: TEMPLATES ∽

The dense body has neurotic perception, then consciousness is trashed even more with food. Fast and all fearful templates dissolve along with the body. As your self-esteem is raised you'll no more look so shoddy. Aging is: magnified templates (from old systems) repeated throughout the life of neurosis. The template prevents adaptation to an ever-changing universe so that with age they are either broken (success) or magnified (degeneration and death). The bad eating patterns combine with this warped and false persona which

125

increasingly becomes a caricature, not a true identity. To the mal-adaptor life seems a big mistake. We want to erase this old take and get to genetic clan—the True Man.

⊰ JOIN THE CLUB ⊱

So you made a mistake: join the club. So you lost your temper—get over it. Shake it off: the past is "the past" because it is "passed." It has no bearing (it's a red herring). Your heart was a-tearing tolerating the overbearing: those so uncaring. Your problem was looking "up to" the oppressors with their swearing, their nostrils a-flaring nor mercies unsparing (their eyes so a-glaring). Don't mind while I'm airing and boldly declaring: let the fast be the armor you're wearing: for wounds repairing and all talents preparing. There is no comparing. Now no more despairing for rid of that past you can be so daring. Seek out the caring—that's a fair pairing, then get ready for profit-sharing.

⊰ TRANSMUTING TRAGEDY TO TRIUMPH ⊱

Dickens transformed his tragic childhood problems into enormous world success. Without these demons he could never have been the writer he was. "I've done my best—let my books do the rest." We can't transmute our early demons into success without self-mastery (of the instincts) without which we become snakes (hurting) or hogs (hoarding). When one is hurt he mal-adapts. It's the mal-adaptation (sin) for which he must repent to break the template to become whole once more. That repetition was a bore—we must close that old door. Though we've been shoved to the floor and made very sore our neuroses flared up in the post-war (and it was sure hard for others to ignore). So just fast and repent then you will surely soar towards all you've yearned for.

Anyone doing something great for mankind is likely alone. The whole human race seems to be marching to a different drummer.

JUST SKIP DINNER

No bummer—we all have needs to be confirmed but the great must just trust God: *He's* their power and comfort--you can feel His Spirit coming through confirming you. Have faith in his Unseen Presence is all you need: Will you lead or be weak-kneed?

❧ BE CAREFUL WITH YOUR WORK ❧

Be careful with your work. However, every writer is different. There are popular novelists who write once and never look back in one book after another. Then there are those like F. Scott Fitzgerald who worked laborious revisions to each chapter in order to achieve the beautiful effects making it seem "as natural as a bird singing." The point is to mature: rid of all neurotic complexities and addictions the True Spirit comes through. Are you blue? You've had a lot of guff to chew having to adapt to those without a clue. It wasn't your due—you were one of the few and treachery was all that you knew. All one's life it was like having the flue: in one ordeal after another it became a rancid stew. Well that is your cue to repent for the future now in full view. Just let it all go and get ready for your debut.

❧ WATCH and WAIT: MASTERFUL INACTIVITY ❧

You must have complete patience as creativity and synchronicity matures underground. This takes great faith in self and God. The Great know the art of "masterful inactivity": just do your work then watch and wait. Once done do nothing and let things happen on their own— that's winning out the gate. This patience is the result after years of persecution for *not* being that way. After years of impatient impulsivity and the tribulation which follows one just wants to enjoy life--like arising early and taking a walk. When

127

things get tough the flesh gives up so now the spirit takes over. One must be in the spirit to persevere, have faith in the unseen and blossom to the sheen—and success. Whether a speaker, writer, poet or painter you must feel you're an ambassador to an almighty King. Going higher than your petty self: this is the revelation you can bring as you make the people sing. To no more habits must you cling nor involve a sordid fling. But like a bell you will ring and suddenly it'll all seem like spring. It can be a terrible thing: blocked talent is like a broken wing, a constant bee-sting or bondage through a nose-ring. But now life takes an upswing (and it's a beautiful thing).

✆ MODERN MAN'S INCAPACITY FOR LEISURE ✆

What makes genius run? Leisure (fun). The true mature genius knows this and makes it most of his life. You cannot force creativity but by relaxing into the receptive mode, out it comes. What gives a writer insights? Quitting work for the day. How to write: Stop writing, just relax and open to nature: up comes a variety of new insights. Modern man has an incapacity for *true* leisure and thus his "fun" is to drink or eat. Enjoy life through your higher senses and then your fun will be your work not the party smirk. Force it without leisure you'll be going berserk. Mr. all-work-and-no-play becomes a jerk—it's like a personality-quirk. No, leisure you must *not* shirk.

✆ LIFE PHASES ✆

Fast, then sail through life. In your twenties you're still learning. In your thirty's you're laying a foundation for the rest of life. In your forties you're accomplishing. In your fifties you're beginning to rule—becoming the head and not the tail. In your sixties you're empowering others while getting into more pleasurable relaxation and less work. In your seventies now you're totally enjoying life. So what's there to fear in aging? Look forward—the worst part was the past.

✆ CULTURAL ROOTS NOT THE SEED SYMBOL ✆

JUST SKIP DINNER

It's hard finding one's place in society when it doesn't fit the traditional role. As a child it doesn't matter—only upon entering the social world do we question such things. I say fast and transcend all categories: attain the spiritual skies. Our cultural or family roots are not the unique seed symbol: the divine blueprint of who we *are* and what we are *supposed* to do. One must transcend gender, culture, race, historical era and family for throughout history cosmic man was always the same: clear of all categories and completely unique. Stuck in past systems your compelled to be meek and the situation is bleak (very un-chic). It's like mis-fitting a clique, being up a creek, feeling a freak even a geek. Well just break-out of the system then become sleek. You'll soon be at your peak and accomplish all that you seek. You've been made very weak (no one listens while you speak) but through the fast you have a beautiful physique and learned technique—you'll now hit a winning streak.

✄ DIS-EUGENICS ✄

In order to realize your true self based in genes you may have to break free from systems showing *dysgenics*: degeneration of all potential through sin. They've let themselves down so cannot recognize true genius in *you*. "A prophet is never known in his own town or family." Let pure God-given talent mixed with desire take you to the top as God gives you all things to enjoy. This is laying hold of eternal life as He richly and ceaselessly provides all things for your enjoyment. So go through the fire, debunk muck and mire for the devil's a liar. You can have it all after such a long haul ever-blocked by a hard wall. Don't you recall? You were made so small. But now you're wearing a prayer shawl and the Lord says "I will increase as you decrease." Fast for peace, hair like fleece (the sign of a leader) while all tensions release. Take your belt in to be a shoe-in--no more the ruin.

Without self-control, human nature degenerates. Study your own dysgenics, the degeneration of your bloodline. Why do whole families fall short of their genetic potential? Sin. If people have no self-control--if they don't make gold at their misery--they just lie down and stay weak.

Remember these types either attack (a snake) or hoard (a hog). Self-control (conquering self) is divine and makes man regal. It's like being righteously legal: rising up as noble as an eagle.

The saints are constant, not cyclical. Repent: *truly change, no more deranged.* Why covet new life without changing your cyclical behavior—bad things done over and over again despite the tragic results and recurrent repentance? Success comes to the spiritually mature—no longer a baby but worthy of prosperity. This is the inheritance of all God's children. The fast links you up to your genes: It's a linking device to the *bloodline blueprint.* In this way you're exhibiting "genetic linkage."

❧ THE TIMING OF THE GREAT: CRISIS ❧

It has been shown by Koestler in *Act of Creation* that true landmark discoveries link with history—something happens bringing the genius into the limelight. As part of nature would this not be true? We all have a unique talent, the seed of a Creative Act inside. But it's a mere *dormant* potential: it must be developed. If pure the Creative Act as part of nature evolves through cycles. It sprouts, moves into place and completes itself naturally. As all of nature's structures it cannot be rushed, but for those with the courage to work and patience to wait the entire act (together with the reward itself) will fall together in a fabric. Timing is all: the minute the act is complete it links with history. Just by doing what comes naturally in each moment, our true genius comes through: we do magical work for magical pay. The test of discovery is: *does it work?* The link of historical events proves it out. This isn't taking advantage of a tragedy but rather doing what God's Champ has been called to do as his destiny makes history. Welcome to genetic body fluids: you've come to your True Self just as you've overcome the world—*because* you overcame yourself.

10

CHAMP RISE UP

Clear Up Cheer Up—
The Inversion of Systems
is About to Occur

No matter what you've done in the past--it isn't you. What you are is your *Destiny*--the Blueprint of all your potentials. You've been held down by sick systems, by cycles of sin and by debris—storage, superfluity, stomach sludge and silly insanities of the slanderers around you. Having seen your obstruction you've dissolved the problem and become a totally new person. Having transformed you've also transfigured--you *look* and *are* a totally new human being.

With your change we come to the gist of this theory: *enantiodromia*: the inversion of systems whereby the top becomes the bottom and the bottom the top. Your time will come if you faint not. While waiting for success, digest the beginning chapters in verse. They will entertain as they edify and enlighten your life (which will go from boredom to a whirlwind of activity soon). Trust me—I had many silent years thinking "it" would never happen. But I hung in there and it did.

∽ UNIQUENESS BRINGS ATTACK ∽

JUST SKIP DINNER

The greatest achievements come from the independent mind. In the preparatory stage (taking decades) he's ever-maligned by the unkind, but at a certain point he comes to light as the Rare Find. Man has a right to exist for his own sake through his own work, for the result is creative genius which revitalizes and elevates his entire culture and then the world. It's a matter of *collectivism vs. individualism*—not in politics but in a man's soul. Unique genius propels the collective--the good, high and noble is only that which maintains its own integrity so that everything comes into Order around *it*.

As one sick family member can degenerate the clan in the *contagion of madness* one realized member can elevate the whole system in the contagion of *enlightenment*--but only by remaining unique. For in adapting to *them* genius is debased, talents encased, destiny erased. If acting out to gain approval he becomes the "bad act" and fame becomes shame. Just be yourself then watch the system invert: Now you're the Grande Gent and Dame in this grand switch: Suddenly the wrinkles leave (it itches) along with human blocks (dark knights and witches) and mostly your sins (they kept you from riches). With your transformation the entire outer world changes too. Now it could be you're ignored by everyone except for *one*: it's that *one* who is a bridge to your new world—you need only please *him*—not "them".

Because genius is larger than life it becomes more misshapen--warped—with food sins or forcing the fit to the herd in which he's seen as "extra" weird. The others conform and "pass go". Genius can't do this so it's kept in the back but with the fast the tables change soon (a castle from a shack). Whereas he feels no one likes or listens to him, suddenly all are loving and receptive. From being rejected and disconfirmed while

133

stuck in a system now he's loved and adored by the whole world. That is the same divisive moment when he ceased eating and

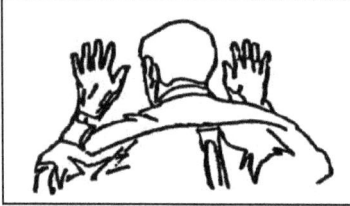

sinning at his lousy circumstances and fasted at them instead. Now he wins all hands down: he's now a shoe-in, no more to ruin—not the tail, but the head.

This breakthrough-uniqueness does not come from religion or atheism but a personal relationship with God, the designer of that Unique Blueprint which is you and your work. But be ready for it brings attack. Here are the points to recall when down:

1. The Big Switch is About to Occur.

Just think: "Switch: it's my time now." Joseph's brothers thought he was dead after they sold him into slavery, but be suddenly switched from prison to palace. That's why you went through so much to prime you for leadership when the tables turn. Just think of that now for you don't have time to be depressed anymore--just prepare to meet God and your destiny. You went through hell to have a war story--to help others up as they go through *their* silent years of sullen salivating slavery.

2. Have Faith in the Unseen.

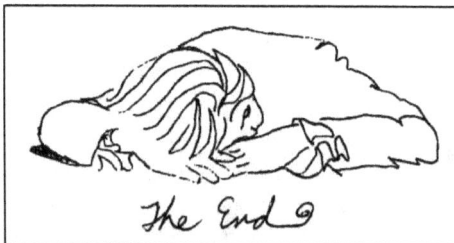

The End

Now ignore all negative situations--how things "seem"--as mere appearance reflecting past thought. Have faith in the *unseen*—that Divine Protective Presence creating your future. Though constantly maturing underground soon the seed is seen as it blossoms into assured success soon --so make room for the bloom (it arrives like a big boom). That's after you use the broom—to clean-sweep cruddy carnal crowd (to their doom). All they did was fume: and for you it meant gloom, but now you've got God your new groom. You've left your tomb for a home in *His* womb so all work can resume. It's like a fantastic perfume or holy costume.

3. Speak to the Mountain.

We're all in some kind of prison: of isolation, despair, chaos. But you're still a Prince. For just like Joseph, the Pharaoh took him from prison to palace: In one moment all can change so be prepared for your pinnacle. Tell death, disease, despair and debt to "wait a minute—I'm about to be made King." For this to occur you must know it *will*: speak to the mountain, it will be moved. People said you were dead but now they'll see you in all your glory instead. This sudden inversion of systems is a very old story. To the extent that your past was gory your future holds Victory.

4. All-Bad Turns to All-Good.

They insult and ridicule: "So you're a Prince huh?" God sees their scorn and always turns it to your blessing. What they meant for evil God turns to good. With the herd whether you spoke or didn't speak you were always in trouble, but now God blesses you in the very presence of your enemies. You may have felt shame as you never fit in. It's far more common than we know and the result is nagging persistent guilt and low self-esteem for life. Stop this strife with self: It's all from being outer-directed--seeking people's opinions or approval rather than the best: fasting and prayer as you spring to your crest. Then

135

you'll say "I may be imprisoned but I'm a Prince because God said it, Amen."

5. Never Let Others Name you.

Let no one tell you what or who you are--never take on their reality *about* you. What you *did* and who you *are* are two different things! Each dawn is a new beginning after repenting of yesterday's sinning. People are filled with guilt and shame—it makes them lame. Just repent (meaning *go back up to the penthouse*) and you're again perfect, the destiny seed symbol from which you came. Let only God tell you who you are. Man's problem is *confusion*: always running from person to person trying to get endorsements, leading to more neurotic reinforcements. The day I gave up people-worship for God and *only* God (the phase of "people-subtractions") and *really came inside* it became a whole new ride. It was like a tidal wave, the easy flowing destiny slide. Having awoke I instantly attracted the right people and it was all downhill, a cosmic glide.

As a transitional exercise, *forget the people from your entire past*: just go to God. To be a success you must delete all distractions—*primarily* petty paltry people problems. We're all gifted at something and your gift makes room for you. What holds us back? Listening to others and taking on their worldview–of you, and all of life too. You must say "I may be lonely (rejected, scorned, belittled) but I am gifted." Let God turn your whole life around--you've been through hell and adversity but now you take center stage. Just wait...it's about to occur, and the whole morbid past will dissolve like a blur.

6. A Prophetic Life Brings Jealousy.

You must play the prince walking in the new Kingdom *before* you possess it. Forget the past and seed towards you destiny—focus on where you're *going*

JUST SKIP DINNER

not the preparatory past when your defects were showing. You must know who you are *before* you're made King--it's prophetic before its reality. As champs we've been through things others can't even imagine. Ignore all trouble as mere appearance and instead take it as proof of your destiny. They're *supposed* to be jealous and upset for the herd sides against the great: so if they hate you, rejoice! You've made mistakes, we all have—but now the past is noise, a reflection of society's false joys (you just lost your poise to be one of the boys). With transformation you're *cosmic*—you like what divinity enjoys and are surrounded by angels who God employs. The real prince is always thrown in a persecution-pit and being in the pit proves you're the prince. As God pulls you out of this whole mess you'll say "God-bless": let *His* glory be your dress when rid of your guests. Despite the trauma they caused you're none the less—you've grown so much by overcoming stress! Now get ready for the press and to God's instructions say "Yes." Stay away from excess (that's *finesse*) and with man no obsess (that's noblesse). Now you're making real progress with no more transgress: this means success, the land to possess.

7. Herd Mentality Kills.

It's the Collective Consciousness—the herd mentality—which spawns racism, sexism and ageism. One really sees this in small towns where the sense of isolation combined with tight group-thought allows disputes and prejudices to grow and everyone to "know". Despite (and because of) this constraint, genius is *forced* to maintain his unique spirit with tenacity. Having gone beyond the townsfolk he finds his own stream awaits, and the universal writer, painter, composer or preacher with a renewed sense of purpose never gets down again—no more parochial chagrin. Can you experience this adventure, where self-contained heroes accomplish great feats? Are you completely and *terribly* alone? Accept it and let God rule, happy you're different from the fool. For being led by the crowd is not cool—despite their nice image they're coldly cruel. This is the greatest experience: wipe the entire past out of consciousness as a whole new reality emerges. Have no more to do with all you're desperate to escape, then wear the winner's cape.

8. Their Wickedness is Unbelief.

Unbelief goes with wickedness—that's why they put you down. They don't encourage from their faith in God, they discourage from their flesh, feeding your fears. Watch who you let influence your belief system! For what we believe determines how we see ourselves and what we achieve. It's the key to success: what we *believe* about it—can you project all thoughts into the future of your glorious destiny and forget the past's phony fair-weather friends and foes of pure villainy? Your beliefs must be strong in God who has *destined* your success planned before your birth. And so genuine success means fulfilling that blueprint--becoming the person God wants and achieving the goals God has purposed. I must believe that God equips me, enables me and is with me despite what the world says (*op-truth*). If you can believe in your success you can *envision* it: have that picture image, then it becomes living reality. That image God implanted *in* you and that's why it's so exciting. Belief is an awesome God-given power for our body and mind can't distinguish between what is true or imagined. Imagine yourself for total success and all of its fancy trappings. Do it now, for self-doubt brings only failure. Just press on to how God sees you accomplishing–it comes from abstaining and dis-acquainting.

9. Self-Confidence is God-Consciousness.

You must die to the past: For God predestined your job, purpose and reward—you can do *all* things through He who strengthens you. He has gifted you with certain spiritual gifts, so why belittle yourself or listen to others who do? When you bow before belittling man you're slamming God, for we are *His* workmanship. People of excellence can simply walk *into* these pre-ordained goals, indwelt by the genius holy spirit and skilled by God. Have confidence for self-doubt is sin: God has made you something special to succeed and achieve. Our troubles just build character to become the person He *wants* us to be. Remember He has promised to *make a way* directing your path, so reprogram your thinking and visualize your divine assets—stamp them indelibly on your mind as the person you want to become. At this last part of the preparatory phase before crossing the Great Divide, you must replace all negative with positive with these words: "God is with me and He's big enough to get me through it all."

10. The Sacred Fast.

The fast will bring you to your True Self and God's total protection on your way to success—health, wealth or svelte. Know this then the creative action of the day increases a hundred-fold! For if your mind's totally on God, out tumbles a mass of miraculous insights about you (past, present and future) and they're all as bright as the sun. Fasting makes you muse over miracles as it solves all riddles.

P.S. Fast Then Walk All Day in Nature.

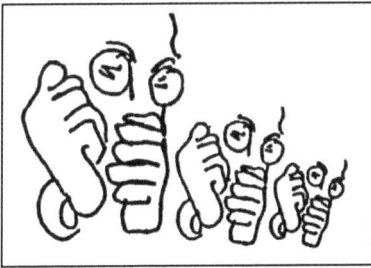

Fast and the light out there is a hundred times brighter, as just *knowing* you're fasting (on people mostly) brings ascension to angels and a much higher state of consciousness. How simple: just to not eat, delete the fleet and stick to the (holy) sweet you gain so many benefits--and how inefficacious therapy, books on psychology or talks with neighbors. People are looking in all the wrong places, as *you* win just by going within. Be a happy faster by just eating once daily. If you have a piece of fruit later just count it as juice (avoid the legalistic fasting noose).

JUST SKIP DINNER

All over this world fasters are coming to the same conclusions: that every moment is a kaleidoscope of meaning unique to each faster and it's *all so good.* He's no social butterfly—he prefers the higher activity of mind and soul. Be a silent witness of the divine as each colorful moment is born and reborn. What difference past mistakes when you can have all of this? They mean nothing—they do not exist. All that exists is this day the Bright Moment We're In.

᪥ BEFORE YOU RISE UP YOU MUST ᪥ OVERCOME!

For the Herd is Obstruction but Genius Overcomes
All by Fasting on People, Habits and Food.

The above descriptions of the outer world and all it's petty tribulations brings us to the last point: before success the tribe gets restless--there is *tribe-ulation.* Overcome your pre-success crisis by fasting, for there are some dark demons that go out no other way. Fast for the week-end in the sacred sabbath fat-fast to pass over to your new realm. With your sudden transformation fasting will be your famous friend from that point on. Hold on--you're about to transcend the throng, for you're right and they're wrong. God speaks to the individual: that's your song. Through the fast you'll distinguish yourself from the mass-- so just rest and wait for the Lord (just hold on).

RX:

᪥ FAST and REPENT FOR ᪥ DIVINE VINDICATION

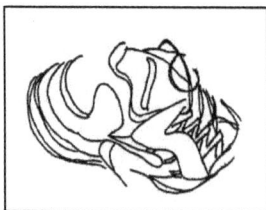

Do you have backstabbing, belittling relatives? Well get ready for the Big Payback. Soon you'll see it as fact for God's revenge is necessary for the relief of the Saints—as proof that God is just (it's like our pact). To think otherwise brings aging cynicism—you rust (you lose tact). Those sting-shots and flip-flops brought constant stress as recurrently your joys did combust. Abuse gave you thorns forming an ugly crust but repentance dissolved these distorted implants from others which blocked success through self-disgust. Recall always our slogan: All men are sinners—your accusers are filled with lusts. See much of your exhaustion as the result of oppression: letting inferiors in to rain on your parade (you got jade) and take the wind out of your sails (God heard your wails).

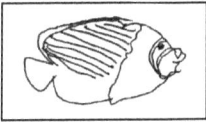

RX: Three days with just water (a drop of lemon in it) OR: the Sacred Sabbath Fat-Fast begins with one protein/fatty meal followed by sixty-hour joyous pain-free fasting.

DURING THIS HEALING PERIOD THINK ON THESE THINGS TO BRING ON THE FLOURISHING OF YOUR "DIVINE PECULIARITIES":

1. *People come and go but the pack personality prevails.* One dies and another instantly reflects the maze-way defining love/hates: The genius (characterized by deviant associations, clear-sighted audacity and austere simplicity) is maligned (people can be so unkind)--but it's all resolved through Fairyland Fasting—our Fancy Find.

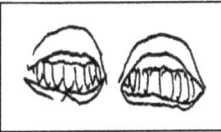

2. *The false churchist loves society*: He sees *being different* as sin yet accepts sin that is socially condoned. Also known as "social hall religion", it's this false piety masking the tribal instinct. But saints in history were loners, spending years alone in the desert yet absorbed in God (having escaped the clod). The lone faster gets power (it's all God's rod).

3. *Championship is like a SNOWPEAK*--the highest point of whitest purity. You get to the pinnacle through fasting and

transcending all society spawning a brand new life. It means no more strife--*today* you can start the happiest time of your life.

Now as you look back at your life you can see how:

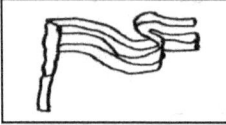

Your OUTER DIRECTION was from the chaos and confusion built on constantly negotiated "meanings" that were just social--not true reality. Stress and insults resulted from interaction with those who saw "consensual" reality as "truth". Insofar as you went along with this your true gut instincts were submerged in order to conform to some norm, not God's form. Now you take center stage—the true sage leading your own forum.

The INNER JOURNEY was your way out: In denying the ego (food and social confirmation) you matured way beyond the silly script (scapegoat status). With just one day on the fast you transcended petty pugnacious politics and came to realize the True Self and God who designed this blueprint.

This was your TRIUMPH as your original ideas achieved universal recognition and revolution while everything came into order around *you*. Out of obscurity and the periphery the superior man came to the center--now who's the sinner? The top became the bottom (boss became a beginner) as you became the winner just by skipping dinner (it meant getting thinner).

P.S. How Vindication Works while fasting:

❧ PALEO: WORLD, FUTURE, DEEP ❧

Fasting is the biggest sacrifice you could make for victory, and God will pay you handsomely. Fasting beauty cannever be denied—it's subliminal recognition is in all humans who see the shiny radiant

saint. Now you'll really look futuristic and paleo: worldly--the best of all cultures. More Latin than Latin, more Italian than Italian. Your abuser will be scared to death of what he did (as he sees others want you) and all your problems will now find magic solutions. Let him drink or be dumb—he won't know what you're up to while you zoom up naturally without saying a thing—you stay mum! Like meeting a *requisite standard* fasting makes you rigorously correct and exact. Avoid inexactitude—be gorgeous and your foe won't believe his eyes while the fast makes you impervious or able to resist (now who's despised).

⤦ HIDDEN TACTICS: ARCHETYPAL INVERSIONS ⤦

Fasting is a hidden tactic to bring on an *archetypal inversion*, so fast for the system and your career. Become black beauty, blond beauty—a fragile porcelain vase—only the worst human could hurt a woman or man like that. Soon you'll have proof so vast so fast to dissolve the obstruction and attain the spiritual skies. For fasting is the only way the weak can overcome the strong: If it weren't for fasting bullies would never be overcome! Being a fasting magic elf is the immediate protection from slugs. No harm will ever come to you while fasting, the lost art without drugs. Even if your neighbors "heard the gossip" just ignore it and they'll find you fascinating. What's a person like you doing in an abusive situation like that? They'll come to your aid and your glory won't fade. Your judgment is now God's judgment on the foe, for you sacrificed and decreased, and he didn't (he has no glow).

⤦ STEP INTO ANOTHER WORLD ⤦

JUST SKIP DINNER

Fasting is something you step *into*—another world for days or weeks. The key is to lock it in: after three days the hunger is gone so no matter what, keep on. You look distinguished like a movie star--something is so different about you suddenly and they'll never guess what it is. You look so universal, an international jet set Creole lady or gentlemen. You're a fine-featured lady of distinction—a scientist, philosopher or poet? Through fasting you will show it. For all this, just take joy in hunger pains—that area is cleaning out, after which things are kicked up a notch (so don't blow it). So to be a star, just be a star. Morals first, success second. The more you fast the more you'll long to step into that world again and again—control at last. Magic cornucopia point to point, in romantic depth and rich beauty—the taste trip can never compare, so happily do your duty (have a blast).

ᘕ FOES COME TO PSYCHIC BLOWS ᘕ
The Way to Win All Wars

When in pain just realize that yoga (meaning yoked to God) is based on the conservation of energy: less physical, more spirit and mental. This ride is inspiring because all energy is now *perceptual*. Sainthood shows so your foes come to blows. Because of the cultural inhibition against violence your foe feels extreme guilt at his brutality because of your new aura-- how you *look*. He can feel something's changed but has no idea what--you look and *are* so radically different. For when you fast all of that 85% of digestive energy now turns on itself (toxicity and fat) and eliminates it. You look at the foe and see all his friends are fat—

you've separated from the low. The system is: you stop eating, he starts eating more and now you're behind the magic door. In the movie *American in Italy* the man said "only while fasting am I the king" with shiny black lustrous eyes and hair. When high on fasting I am the desert queen and my victory is won for the foe has no sheen—for only that marks the genius gene, the true scene, the teen, the keen, the pristine, the toughest marine. Just by not-eating we're in another world like a luxurious submarine. The foe seems ten years older while we've de-aged while getting bolder.

11

RED:
The Paleo Path

Letters of the Fruit-Fat-Fastarian Chieftess

I'm speaking to the masses steeped in grains--what a strain! You're not to blame it's just your taste needing to tame (from the yeast, it's a beast at the least). It's your time: no more blind leading the blind. Am I unkind? Well hear my mind: Eat your fruit and fat but take no advice from mice. They act nice but eating rice it irritates like lice. This food-type brings spite not might, so see the light: eat right and you'll be light. Watch what you bite: the starch makes you fight not increase your soul-height. Just eat fruit or fat then fast and let your insights alight (ending this blight) and start you flight. No more will they slight the dignified knight (you'll be that tonight). Whether you write or just in the right you'll be a good sight: pores so tight not puffy skin so white.

✺ REVERSAL DIETING IS NEOLITHIC ✺

Higher Paleo-Fasting is feeding on eternal things for man—fruit, fat-fasting, the elements (sun-wind-stars at night) and God--while avoiding temporal distractions (people, habits) and also (modern man-made) food inventions—pasta, bread, cake. To return to our paleo-beginnings is actually the revitalization of culture, the

146

Renaissance of Man. The real diet of frugal fruit alternating with fat and fasting promises a release of total creativity and mind expansion on earth. Culture reshapes as man finds himself— because his brain is finally fruit-cleansed and fat-fed. Our primitive ancestors ate fauna (animal protein and fat) but when they didn't make a kill they just ate berries for weeks and sometimes they didn't even have that—they had to fast. This is the true meaning of *au natural* : reversal dieting, the marriage of Atkins (fat) and Ehret (fruit and fasting). *Bon appetite!* The caveman ate some meat then some sweet. It's a great treat: the food elite avoiding the wheat. But don't take a backseat: only fasting makes you tough like concrete. Like hearing a small bird "tweet" your senses become replete and there's no defeat. Take a retreat: Fast, fruit and fat to reverse and it's the coffin you'll cheat and you'll cheer all whom you greet and what you excrete is *so neat.* In foods be discrete so you can compete while your work completes. It's all in *what you delete* to reach heaven's suite—I'm talking easy street so don't get cold feet.

FAT IS BEST, STARCH IS WORST

Drinking water does nothing to moisten the skin and cells. It is the metabolism of fat which creates water in the cells. We need fat-- esp animal fat—not gallons of useless water which is just compensation for dead dry food.

Fruit reversing with fat and then daily fasting is the "higher paleo" diet. But there are two prejudices we must dispel: that against fat and the fear of fruit-sugar. Animal fat is not the problem refined starch is. In 1900 butter consumption was 18 lb per person a year and now it's only 5 lbs, yet heart disease was rare then and the number one cause of disease now. In 1900 40% calories came from dairy fats yet heart disease and cancer was rare until 1928—when refined oils, sugar and flour were refined *en masse.* Low fat high-carb Hindus die from heart disease as much or more than meat-eating countries. Eskimos have low heart disease but a very high fat diet. The French are high in saturated fat but low heart disease.

CHOLESTEROL FEARS

It is refined starches and sugars that increases the insulin which converts food to body fat and cues the liver to create cholesterol,

147

not animal fats. Fats elevate glucagons which puts the body into fat-burning mode, dilates all the vessels and airways and stops the building of arterial plaque. The problem is starch, not fat. Fruitarians worrying about avocados and olive oil are just taking on the mass mythology surrounding fats. Fats are ideal. Starches comes through the skin pores creating puffy white wrinkles, while fat brings water to the surface to moisten the skin instead. One of the best results is the predictable regularity--as the colon needs fat to work properly—which simultaneously boosts energy to the height. It happens overnight and the feelings are out of sight.

❧ FROM FAT TO FRUIT ❧

In a completely clean system a little fruit sugar like one fig is all one needs to sustain the whole day. Until that point of cleanliness sweet fruit can bring on a rapid detox and cause problems. It's best to stay with non-sweet fruit in the transition. What I'm calling for is an Italian diet: a few grapes, a little cheese and how about some fish if you please. In a clean body starches come right through the skin pores, the whole body looks white and chalky and a depression sets in. Refined carbs have a high glycemic index so they should be totally avoided but fruit-sugar carbs like oranges— although high in carbs—have a low glycemic index and should be consumed. It's "in" to be down on all sugar even fruit—while continuing to consume grains, a much worse form of "sugar." Starches lower HGH and the result is failure to thrive just at the age when one should be thriving, for at his highest peak he should be at the top of his game, not brain-body-system-lame immersed in the eating game hanging out with gluttons same.

❧ GRAPE CARBON IS THE EHRETIST ❧

It's wonderful to be grape or fig-fasting in the mornings then some fat like avocado or doughless pizza to reverse gears. Then fast to prevent tears and lengthen your years. The benefits of the grapecure are amazing particularly after a diet of wrong food. Without being crude it's the fragrance we exude and it really sweetens the mood. Not that we obtrude but let me allude to something you should include. The fruit-fast will preclude your inner feud or looking like a craggy dude. Now your cycle can conclude and you'll even look good in the nude.

JUST SKIP DINNER

One can exist on fruit and fat for life without the devastating effects seen in longterm "gluttarian" fruitarians who binge all day on sweet fruit: constriction, bloat, rotted teeth, hanging/wrinkled skin, mood swings, irascible cravings, isolation depression, fatigue or insomnia. They eat too much without using fat to offset the deficiencies of pure fruitarianism combined with the daily fast to build health and character.   How about some cantaloupe then a piece of fish? Forget about supposed "food combining" rules— they encourage mono-dieting and it makes us into fools. For when I replaced fruit-gluttony with mini-fasting after fat and some fruit juice I felt much better and wasn't hungry all the time (a sign of a constantly-detoxing gut).   Fasting from noon to noon (the "glory church") is simple after a nice FF meal. Happiness comes from the utter simplicity of eating this way—no more truckloads of fragile fruits to store causing emotional flip-flops galore. It was the appetite-suppression that   thrilled my soul so I could just concentrate on my goals and so much more.

❧ FRUIT: CAUTION UNTIL CLEAN ❧

Sugar--even fruit--is volatile for someone who is highly toxic or has hyperinsulinism. Many of my readers were miserable with symptoms and craving until they removed sweet fruit entirely for the interim of the transition, sticking to non-sweet fruit. As insulin problems recede one can enjoy true fruitarianism: frugal sweet fruit and fat which will sustain you for twenty hours a day.   With just fruit, detox can be heavy so one must self-regulate the flow with fat to slow it down. Thus fear of fruit sugar is not totally without reason-- it's a matter of the season: when you get like Jackie Gleason stick with more fat than fruit for it one is toxic it will be treason. After a time of eating just fruit and fat with daily fasting the weight begins to strip off and you'll be adding more fresh fruit to experience an incredible high.   Once properly nourished the fruit-fat-fasting diet has an unbelievable effect on consciousness, skin , respiratory, head shape, vitality, ideas, dispersion of negativity and favor in public.

❧ MAXIMUM MEAL MAP ❧

This is what I've found to be a superior schedule or "maximum meal map" for two types of daily fasting:

149

JUST SKIP DINNER

✦ BUDDHA FAST ✦

Morning drink like coffee, apple or grape juice then a little peanut butter or for paleo-dieters doughless pizza for lunch. Now just skip dinner.

✦ RAMADAN FAST ✦

Breakfast: Morning drink, doughless pizza with cashews.
Dinner: Raisins and nuts or pineapple chunks.

✦ SACRED SABATH FAT-FAST ✦

Saturday breakfast: Cheese Spanish Omelet with Guacamole followed by fasting to Sunday night.

In the MMM (maximum meal map) the fruit dissolves all albumin and mucus while the fat lunch releases HGH, creates energy, suppresses appetite and releases all water from the system. It's the big streamline, then the 18-hour water fast releases more HGH. The MMM is the "maximum youthifier". The following letters refer to the lacto-fruitarian bliss of the natural man--the champion not the Peter Pan:

FRUIT- FAT-FASTARIANISM
Letters from FFF Chieftess

✦TRUE ROYALTY✦

JUST SKIP DINNER

True royalty is identified by eating habits. Fruit and fat is the diet of nobles and fasters are the elites. The superior diet of Kings--both fruit and fat--has high water content in contrast to dry beans, rice, bread, flour and cornmeal which are only 5-10% water reflected in either dry wrinkled skin or fat. But beautifully rich fruit and fat circulates through each cell giving superior strength, endurance and power marking the elites. A fruit and fat fed bloodstream dissolves impurities as it rebuilds until perfection. Excellence in all things results. The high-water diet of *Vitarianism* means "cult of life ennoblement". Fruit and fat-fasting releases energy and this means jubilance. This creates boldness to hit the bull's eye every time, for the perfect food of man *perfects* thought and action. The King's Food is fasting, fat and fruit--not mono-produce or cake, rice and bread. The former creates elegance—the latter makes one puffy and gruffy. The whole raw vegan movement (being based on false dogma) has become stuffy.

Purified and paleo-fed the old life is gone as the new one bursts open. Now everything converts to its opposite: the shift from pauper to prince, slave to king, jail to palace. Rising up in attractions is the miracle of fruit-fat-fasting, God's foods for man. Knowing it's the paleofasting keeps you humble and in humility lies Great Power. To succeed: stay with real food then just be sweet!

Eat Italian without the Starch. Masses eat cake or corny concoctions while nobles eat *real* food from the beginning of time: some grapes, a little cheese and fish if you please. Fruit and fat followed by fasting gives the edge to stand strong under pressure and be firm stating facts--it's the difference between ineffectual life ending in horrible death and happy long life marked by golden productivity.

Red-Skinned Forever Dr. Marikelok

❧ BAD FOOD AND CHARACTER ❧

JUST SKIP DINNER

Dirty or anemic blood makes for bad temperaments—weak and touchy personalities. As all systems irritate it's like possession of another force—the non-self: a fraudulent invader, a pugnacious prevaricator. The dual-mindedness of fat-free starchy food addicts flips from desire to queasy aftershock. Caked-up cells house an angry spirit as the false self prevails and the True Self subsides. The caked neurotic splits between *who he is* and who he *says* he is—his social image. With anemic blood and dense mind the caked kook relies on image to cover his vicious changes of mood coming from self-divorce. This inner conflict explodes: anger is a bio-device temporarily resolving this tension from being *two people in one*. It's a case of a wolf in sheep's clothing—watch out for the "nice" with the character of lice.

All psychology comes down to the proper foods combined with fasting. In fact, forget psychology—just eat raw fruits, fauna fat and then fast each day. The everyday charlatan switches from "nice" to not-so-nice as promiscuous and continuous eating enrages the system. But this is normal for the carnal crooked culture which reserves pugnacity for the fastarian outsiders, thus preserving and validating the addiction: the continuous-eating consensus.

This battle between true and false forces one to build strength in the face of this pandemic resistance to the correct paleo diet of fruit, fat and fasting. It's so prevalent we can practice every day. Expecting their vicious vegetarian low fat moodiness readies you to stay sweet as you stand strong. Never force the fit to them—just force them *up*.

Dr. Marikelok iron-red psychology

152

JUST SKIP DINNER

When things get hot do you withdraw from strife? Sometimes we have to stay and fight. Someone has cancer and needs the Grapecure and your pasty fat relatives need the fat and freedom from flour fakery. Though blocked it's urgent to advance the info now. Don't let government "consensus panels" determine how you feed your children. As medical missionaries the tactics we use to fight ignorance determines survival or death. It is sugar and starch creating degenerative diseases, not fat and fruit fasting. That's the opposite to how the herd thinks so get real and promote the paleo revolution: Are you strong enough to stand up and say "fat fights heart disease" and that "starch and sugar creates it, along with rogue cell proliferation which is cancer"?

Stand strong: Expect tribulation and use it for practice. Buckle with stress, you need more work. The Fat-Fruit-Fastarian path is a boot camp to increase strength. Only the greatest can divide from the mass fat-phobic view—one man *can* be a majority. Eating right is just the beginning--now you become great in other ways. Real persecution about the facts occurs most in families, churches and workplace where thoughts that "fat is bad" is seen as self-evident truth. Stay close to higher paleo-fasting friends and receive necessary nurturance there.

Only by building warrior strength can we help the weak, sick, children and animals. We can't succeed whimpering, wounded and withdrawn from the vital war of viewpoints. With perfect body we live painless and powerful—so let's cash-in on the crooked false dogma and caked carnal conflict and change culture. Think of cancer so easily cured through grape-fasting or the anorexic perfected in three days of eating frugal fat. Think of the universal

153

reactor who can only survive with pure fauna—yet is always mis-advised to do the opposite, namely eliminate fauna entirely and include yeasty grains! I tell you these quack theories are constant rains. Bearing the truth we must now confront the destructive and fatal advice of fat-phobic physicians putting us in chains.

Dr. Marikelok. just get red: eat real.

✺ MENTAL DARTS ✺

The real separate from the crass conformity of cake and cookie culture. It is "white": filled with reactive mucus they are pasty, puffy, brain-dead and dry from low fat lifeless foods—and it's all covered over with a whitewashed image. Bound by the low fat tradition it persecutes the dark and different--the gypsy hunter and gatherer--which we are by living on sunlight through fasting, sun's kitchen in the fruit and nut trees and the pure iron punch of fauna protein in between.

As paleofasters we have a new view of reality—the life of the glucagon-elevated mind and body as it energetically connects to the universal source, eternity and destiny. Finally clear from the correct nutrition combined with daily fasting we become filled with energy enthusiasm and excitement. Things obvious to us are hidden to cake and cookie culture kooks who are divorced from the body and particularly the mind of the True Self. The fat-fruitarian "gypsy" lives a separate reality from the mass. He may be called paranoid, perfectionist, vain and arrogant for deliberately setting himself apart from the fat-phobic-obese masses who think they know the truth, while degenerative disease grows by leaps and bounds.

JUST SKIP DINNER

The paleofasting body mystic has psychic attunement to the environment with the fat-fed force of a biological instinct. He looks and feels Indian as the iron red blood shines through the clear sunned skin and as he relates to nature, God and the True Self (now revealed). The paleo-fasting path is the quickest way to this unique design and the rare genius therein. I once thought pure fruitarianism made me psychic and sun-loving but only by adding fauna fat could I bike all day in the sun and become really intuition-smart with clairvoyant and déjà vu experiences. With fauna fat in the diet I felt as though I was riding on a wave of perfect synchrony but with no fat I was "grating" not humming. The fruit meal cleaned to the unique design but the blueprint of all gifts and talents as prefigured in the seed emerged from the fat-fasting. Longterm fruitarianism is fallacy--you must alternate with fat and fasting to have a spring in your step while your light emerges with real pep.

Caked dry society may not admire you for your endless endurance, miraculous mind, unparalleled performances, in-depth inspirations or constant cheer. So what? Just know your*self* who you are—truly free and ripe for achievement for you've found the right higher paleo-fasting way which humans have adapted to for 2.3 million years. The fruit-only diet makes you an unripe fruit while fat makes you ripe and bursting with continuous energy to boot.

Dr. Marikelok really red today

❧ JUST STAY REAL AND SPEAK ❧

155

JUST SKIP DINNER

Low-fat phony food is the biggest issue dividing the human race. It transcends gender, race, religion or age. The culture eats empty cloggers while we eat powerfully real paleo, then fast. What a divide! The false food builds such inferior cells it's a miracle they're living at all, so it's no surprise the prejudice--if one can't think clearly "anti-fat dogma" becomes a knee-jerk reaction that requires no thought. Anti-fat is pure phobic dogma reinforced by the growing obesity around us. Does it ever occur to anyone that obesity gets worse as the anti-fat dogma grows? No--for without a fat-fed brain no one can think independently: the dense mind just naturally falls in with the majority view. While dry vegan, starchy and sugary foods bring the degeneration of mankind, *real food* ▢ with fasting leads to its quick evolution. The paleofaster is a *hyperevolute*, a *neanthrope*, a New Man resurrected from the past. Being properly proteinized he's a powerful yet gentle, ageless child--not an empty image of "adult" split from the body, densely dumbed down and dull-witted (though computer literate). Without strong nerves, glands, heart and brain there's no *fidelity*—wide perception of the lush variation in the moment. The dense mind must learn by rote, not receptivity, for the improperly-fueled brain degrades and even deranges. The dense stick to legalistic concepts not the obvious truth in a flower. They want power but due to wrong food end in a cower. Eat right and blessings and creative revelations will shower.

Give up continuous eating of wrong food and your symptoms leave. Give it all up and your sin-cravings dissolve. All vices are just devices to self-medicate the constant dull drag from dammed-up circulation. Would you put bricks and sand in a Rolls-Royce engine? These modern dry foods lead to rot. The body requires what it requires--whether you like it (and agree with it) or not. Instead of refusing to eat the right food because of the cruelty in the meat industry you should eat right--and then garner the strength to insist they do things humanely. Otherwise you stay brain-dead, belittled and beaten--you fail to thrive. The need to self-medicate or spark a dead horse leaves when we choose the *higher* delights felt by elite engines running on correct nutrition.

Dr. Marikelok get red: stay real.

ೊ WHITE SOCIETY IS CAKE AND IMAGE ೊ

JUST SKIP DINNER

"White" society is pasty blockage and lopsidedness: mucus-filled and dense with a whitewashed image to veil the endless mood-swings from the low fat cake and cookie culture. Society is convention, tradition and ritual and all social life combines with food and food fallacies (they eat raw foods but still have dark moods). Clogged cake culture is a pecking-order system filled with status-tension (who's superior to who). As if that's not enough deadly disease looms large—obesity, cancer, heart, aging, treachery, lies and crime all from non-paleo food inventions and fallacious food plans from junk science.

Our children crave fruit, fat and sometimes fasting but the false system tracks them away to join the ranks of the concoction-clogged: the false self trapped in the false body built by giant dumb cells created by non-evolutionary foods. The inefficient and superfluous flesh is then covered with uncomfortable fashions to be "seen". This corpulent carriage compels an *image* to confirm one's value. If one can't think or operate efficiently his unhappiness scapegoats (degrades) others. How to be unconditionally loving and accepting: become the True Self and be happy through the proper diet combined with daily fasting. There can be no "love" by eating from "theory" rather than the true reality of the body's needs from the beginning of time.

The anorexic seeks to avoid the false flesh but then is forced to feed on low fat fake foods of fat social society. For this poor girl it's no fun, just more humdrum. Her only answer is to release the boring and mundane social world and bask in the lights of ageless cornucopia from fasting after eating fruit and fat. Now you're in a happy colorful land where only childlike saints and true genius play. Permanently youthful fasters are busily creative and crave-free until the moment of death.

157

JUST SKIP DINNER

Dr. Marikelok red and ready for revolution

∽ ATTACKS ON PALEO-FASTING GYPSIES ∽

It hurts being the only paleo-fasting reclusive gypsy stuck in family or group. Though by refusing to eat their concoctions and breads he gets well, the better he gets the more they sneer. The paleo-faster may feel shame until he takes a holistic view of the great differences between the *dry anemic* and *real rich* bloodstreams: The caked low fat-foodist is so busy battling physical discomfort he socially scapegoats (shifts blame) to resolve stress and not face his food-addiction which creates constantly-elevated insulin evoked by sugar and starch. To him we're just too thin, too high, too fast. It's called ENERGY which we have and he doesn't.

It helped me to know these subtle attacks were universal to all gypsy hunter-gatherers. We attract these strong reactions because a fasting mind coupled with a strong red bloodstream—the omnipotent eternal supreme varicusa—is so different on deep primal levels. We simply attend to different stimuli—we are "fine". The properly-fueled brain has attention to detail while the dense mind lives in the world of superficial understandings, knee-jerk reactions and science-through-majority-rule. The truth is: fat makes you thin and the deletion of fat (substituted with starch) makes you fat.

I went years on just fruit and collapsed into depression and despair. When I ate animal protein and fat I bounced right back. I lost all hunger yet was filled with elation and enthusiasm as I was able to fast 36 hours between meals. It was like emerging from a dark tunnel and I was fired with desire to bring a fat-white-anemic-fatigued and falsely-led (by junk science) culture back to real reality: higher paleofasting bliss not starchy idealistic vegan dogma. Eat your protein/fat--*give your dogs protein/fat*--and with your new energy enthusiastically reform the meat-industry. It can be done. You will now have the strength to withstand ridicule from a flakey culture trying to make you look ridiculous.

JUST SKIP DINNER

After feeling strange in a strange land I was greatly relieved to know that all gypsies sense subtle shirks. Defenses dissolved I relaxed into my own creative stream. With attacks I can love the poor sick souls still stuck in silly food theories making them fat and sick. With persecution the paleofasting redskin defense is dignity, love and insulation: *aristocratic reserve.* Fasting pulls the head into elongation (*encephalization*) and this marks the monarch. With new mental vistas the fasting monarch is the natural Man, not herd adaptant fed by "natural foods" like huge salads, cooked veggies and starches in patterns of constant eating.

Dr. Marikelok: red and relieved because I eat real.

∾ COLLECTIVE INTUITION ∾

There's never a need for competition. The new Paleo man flourishes in his *own* stream where there is none. God's man competes with no one—he just does his own thing as prefigured in the seed which has finally emerged from eating right.   As each path is unique so is each part we play in this mega world revolution and solution to world starvation and disease--going *higher paleo.* As our power base grows we'll push to replace the old guard with the new blood. We'll plant dry date palms and nut trees all over Africa and make schools, hospitals and prisons REAL with fruit and fat. The institution diet of starches makes people worse than they ever were going in.   We'll help a race spinning down to hell (hideous death in a hospital)—we're destined to do this because we have witnessed the miraculous results of paleo-fasting on our own bodies and minds.

Real, red and regal in the desert sun, Dr. Marikelok. Real food begins the revitalization movement in science and medicine as M.D.s

159

and Ph.D.s come back to life and can finally think clearly after the starchy vegan onslought of the last thirty years.

✌ THE NEW RENAISSANCE ✌

Did you know this paleo-fasting revolution is the Renaissance of Man, a Revitalization of culture? It promises a release of total creativity and mind expansion on earth as all men become their True Selves--each with a genius all his own with no sense of (depressing) competition or sterile hierarchies. Clean the body and the blood, and culture reshapes as Man finds Himself. The millennium marks the battle of minds: society's implants vs. unique genius-now-revealed—because now *God* is his "guru" not some phony man. Only as paleofasters can we succeed despite the overwhelming odds. The physical vessel perfected, Man blossoms like a peacock.

We are all born as unique seed symbols containing all of our potentials. These multi-talents lay dormant until the blood and brain is corrected. Creativity is not something we *have* but something we are *open to* as energy is perfectly conducted via synaptic clarity and proper blood composition. Lose dead weight from caked concoctions to reveal the fine elegant masterpiece fed by the supreme varicusa. It's the blood! Its from the real food from the beginning: fruit and fat!

Now is the time for the new elite to come to the center out of obscurity and the periphery. The new medical model points to the correct food and fasting doing the healing. And the healers of the future must show perfection—the effects of the perfect cure—in themselves. But this new elite is still held down by the dry-dense-in-power. This is a call for a medical revolution.

JUST SKIP DINNER

Marikelok: Forget everything you know—just eat real and fast.

✺ INNER JOURNEY TO GENIUS ✺

To be real and red is to be *clear and creative*. Man at his highest seeks the True Self and this is the Creative Act. Just eat right and fast, then open to the vast opportunities alighting your world. Your magnetic attraction intensifies and a beautiful new vista opens before you—your glorious future. Every paleofaster has a unique beauty. Now revealed this beautiful countenance opens doors. A perfect paleofasting body reflects as pure character etched on the face (a diagnostic picture of the whole). The red paleo path cannot be faked. Bring out the True Seed and you'll recognize the eternal symbol--how God designed you from the beginning.

As fruit de-obstructs the throat and fat-fasting brings you personal power a beautiful voice is yours to express the truth along with strong nerves to stand collision with the crooked cake and cookie culture. You'll now have courage to confront society's biggest plugs: med, drug, food, school and jail. You'll become strong and versatile balancing these old deep factions around you. Whatever the obstacle, bring it on!

Allow the creative process to explore the circuitous uncharted path to the unique self—the raw talent of creative genius. Your self is your Creative Act. Be an artist of self, a scientist of synchronicity: Expect miracles every day, every hour.

Red from sun, fruit, fat and fasting   Dr. Marikelok

✺LESS IS MORE ✺

JUST SKIP DINNER

(peel to the core)

As the body peels off layers of false flesh built by unreal food inventions the *outer idiotic world grows as menace.* A fat culture hates the thin. We know and Einstein proved that *less is more*: *most* creative energy comes in *least* mass. But try to thin and you meet a bank of hostility and hate. The anorexic seeing the light, or the solitary gypsy fruit and fat-faster may feel lost in a sea of starch and sugar.

✍ World Spa: ✍

GLORY CHURCH

JUST SKIP DINNER

Dedicated to igniting the creative spark in man through proper diet fasting and the solitude eliciting True Genius.

✍ THE FLOURISHING OF PECULIARITY ✍

The new era of the hyper-evolute Renaissance Man is the *flourishing of peculiarity.* The only way to become "peculiar"-- which is True Genius--is to properly feed the brain, heart and glands with what it needs: the higher-paleo diet of fruit and fat combined with fasting. As we clean out the old and rebuild with the new foods we chisel down to our core—more deep, more unique: notice the sameness of caked conformist culture! Underneath layers lies the blueprint of "raw" talent never seen before: *Old World Exotic Design in High-Tech.*

JUST SKIP DINNER

This design if still eating too much may seem odd and strange. But the paleo-fasting genius soars to the heights of potential taking the high prize. Fruit, fat and fasting is the break-fast of champions. Having to deal with the blows from hostile human nature is the obstacle which when overcome brings this fancy flowering.

Living my own life in the red and real Dr. Marikelok

THIS IS A CALL FOR A NEW SCIENCE BASED ON EINSTEIN'S THEORIES OF ENERGY WITH A GENERAL FORMULA FOR ALL LIFE SCIENCES:

Bliss is: Freedom from obstruction

JUST SKIP DINNER

All success is attraction
All disease is obstruction
All recovery is elimination

Karen Kellock Ph.D.

GRAPECURE FOR CANCER BUILD-UP

Energy really does come down to grape carbon, just like Ehret said. I always have raisins on hand: if I have gut-pain from hunger the raisins instantly bind the mucus-irritant twice as fast as a juicy fruit. They are also filled with iron, potassium and magnesium—essentials for feeling strong. Johanna Brandt: *The GrapeCure for Cancer: Cure and Prevention*

The grapecure is in three stages:

1. Complete dissolving of the encumbrance.
2. Complete elimination of poison.
3. Complete rebuilding of new tissue.

This *thorough* regeneration is what marks the grape. Cancer cannot be cured through fasting since it checks the growth but it is again strengthened when food is re-ingested. The grape eliminates the cancer, curing entirely. It is the "Queen of the Fruits" with the highest dissolving action and the most complete nutrient. Man can live on the grape for weeks as the amino acids become proteins plus all needed nutrients.

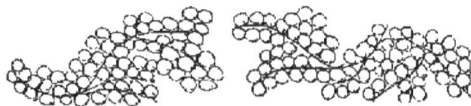

All abnormal growths and tumors are dissolved by this powerful chemical reaction. The cure is mentioned throughout the Bible: it is a divinely simple gift to suffering humanity. Often used as a last resort the grapecure shows complete recoveries despite hours to live. The patient must have faith a strong desire to live and the network must encourage the healing process.

JUST SKIP DINNER

Stage One: Dissolving of Encumbrance. This is grape-only (juice grapes or raisins) every two hours (8 am to 8 pm). This stage takes 7-10 days and it is here that painful symptoms may occur. The patient should take joy in these points of pain since it indicates that cleansing is occurring. Until the encumbrance is dissolved the real relief does not begin. The grape does not stop until all is eliminated. The now-stirred-up poisons must come out lest they auto-intoxicate the blood again so one to two enemas must be given daily if no bowel movements occur.

Stage Two: Elimination of Poison. Grapes will be combined with other fruits but not at the same meal: grapes (8 am) fruit (10am) etc. Having eliminated the biggest chunks, deep cellular cleansing now begins. Old pus poisons, prescriptions and foods from puberty release from cells and dump into the blood. Cell-cleansing is the most intense phase of the cure.

Stage Three: Rebuilding of New Tissue. A complete regeneration of all organs. The salt balance is restored with tomato, red bell pepper, cucumber, lemon and olive. This "fruit salad" at noon is preceded with juicy fruit in the morning and figs or raisins at night. Daily elimination should now work perfectly.

The worst is over—each day shows more beauty and health. The ratty cellulite—coarseness and wrinkles of the skin (from debris storage) becomes baby-fine moist smoothness as debris is broken up and dispersed. Perfection is achieved!

JUST SKIP DINNER

The most common mistake of the medical profession: The patient cannot eat—injurious fasting is followed by eating solid foods "for nourishment". This process is followed by an eruption of the cancer which was arrested through fasting but then gathers stronger hold with eating.

One Science of Metabolism: There are *two*—not one—sciences of metabolism. One is Ehret and the other is Atkins. Let's discuss the former: The human body is an elastic pipes-and-tubes system of 62,000 miles of fine capillary with an inner tissue system resembling a sponge. Anything but fruits and fauna will not digest properly so nature adapts by storing it in this sponge system. The average person is carrying around twenty pounds of morbid matter—debris from head to toe. It is mucus, feces, pus, poison deposits, morbid watery flesh and decomposing tissue! This awful mess putrefies into "disease". The specific 400 disease names refer only to the *specific* point of encumbrance. Modern man is a walking cesspool as the entire pipe system is encrusted—and 70% of autopsies show decades-old feces and worms.

REVERSAL DIETING

For optimal results, alternate fat-fasts with fruit-fasts. The grapecure is a chemical action working amazing results and you'll have more energy and strength than you've ever experienced. The grapecure's like a bell of the highest ring. It'll thrills your soul like a bird chorus sings. The delicious beautiful grape is the Diet of Kings. Try it, you'll love it--it's like floating on wings as your senses alight like a heavenly symphony of strings. So many such delights above all worldly things. You won't believe the bliss this all brings!

FASTING

167

JUST SKIP DINNER

So you slipped? You intended to fast only to end up eating all day?   Relax. Just take note how boring it all was, how you got nothing done, how self-disgust over the lack of self-control ruined your day. If you can just do that you've benefited greatly from this day, so never forget it, I say. Now take note how you felt on fasting days and you've won—never having to experience this again, as new life has begun, amen.

✥DISTRACTION ✥

Right before triumph the devil distracts through confusion. We fail only through broken focus in this *pre-success crisis*. There is a purpose in this chaotic storm—as trouble teaches patience and perseverance it's also a signpost of the good that's soon to come: having learned the lesson of *system inversion* we've automatically won. The worse your problems, the bigger your destiny! Know this and you'll again become friendly. Just when you feel crazy with critical mass there's a turnaround to good. Then you're rewarded for what you endured or what you cut loose (the hood). Crisis makes you prune your life of toxic relationships and other obstructions as lifeless as wood.

✥ IRKSOME TASKS ✥

Whatever the irksome task this is the best way to maximize the results: eat right then fast all day before. If you didn't fast, postpone the task one more day: this makes all the difference as you become happy and quick. Learn to just skip dinner and make it stick. It stops the enemy's trick just as you're no more thick. You find all solutions in clarity, the opposite to sick. Whatever the problem to be solved, just fast and wait 24 hours—it'll all be dissolved.

✥GRAPE THERAPY✥

JUST SKIP DINNER

As it turns to fall 2006 we're in a new station. The old life was constant disputes but now I've returned to my Italian roots so bring out the harps and flutes: I prefer the European route--my old grassroots (to the hour and the day it salutes). This page is dedicated to the age-old GRAPE CURE combined with the HIGH-FAT DIET.

Due to low fat propaganda millions of people drink oceans of water to cure their dry skin--it cannot be done. Only the *metabolization of fats* creates water in the cells. Because he needs fats and only achieves perfection through a high-fat diet (esp. saturated animal fat) these fats are delicious to man.

If you don't want meat just enjoy fish (rasta) dairy (lacto) or eggs (ovo)- fruitarianism. Those on lowfat diets end up diseased, fatigued, weak, dry, wrinkled, hungry and angry. The high fat diet along with the grape cure lends superior health—the high life free of strife. Because God made fats and grape carbon for man, we rejuvenate as God's wife. There are many vegans who may agree with the need for fat but attempt to get it through non-animal sources. They cannot—B12 deficiency is their common lot.

Karen Maria: What about the raw lifestyle exploding all over the internet?

"More herd dogma. Superior species (eagle, lion, man) are *oligiphagous*--existing on *fewest* varieties not all the exotic things on the earth. Sure we enjoy salads and raw fruit but why make a religion out of it? You're just giving into the rawfood seller-yellers who are making millions. Junk science says we get B12 from plants—ignore their rant (it's just a chant).

⊰*Celebrate the Seasons of the Day*⊰

JUST SKIP DINNER

with the Grapecure. Energy really does come down to Grape Carbon, Just Like Ehret Said. Here's the routine maintaining the sheen:

❧MORNING MUSIC: THINK❧

First you arise early, before dawn. Due to diet and lifestyle, it's simple (not even a yawn). You just jump out of bed, earnest for the day ahead (getting out the lead). The mornings are for coffee, being with God, a morning walk in nature, editing your work and just puttering. This joyous morning is without complaint (no bitter muttering) as the happiest time of day, next to the next phase: the mid-day bouquet.

❧LUNCHEON MUSIC: DINER❧

Pets Like Times of Day--They Love it When You're in the Ray

Main Meal and Grapecure. This is the time where you stop all work and start the noon luncheon, the main meal. You enjoy it for three hours, just stepping back from your life and surveying the whole--seeing all your life in a bowl. This is the time God console's--and you thank Him for blessings in your new role as a blessed soul on the Grape Therapy Stroll. *Warning: if you prefer wine to unfermented grape, do not take ANY medications even aspirin with it: it could miscombine and you could radically decline.*

❧AFT BLAST MUSIC: PARTY❧

Conviviality in the Sun

Aft-Fast Blast: You've drank and eaten, it was a great time unbeaten. but now is the time to maximally sweeten--the aft fast blast: *apres* the luncheon period comes the miracle myriad.

JUST SKIP DINNER

ᚼ𝒜𝐹𝒯-𝐹𝒜𝒮𝒯-ℬℒ𝒜𝒮𝒯ᚼ

STOP FOOD/DRINK THEN JUST SKIP DINNER

Yes grape therapy is a gift from God and it could do your health good. In Europe the wine is seen as holy water, a medicine without which you'll die (that's no lie). The bible seems also to be pro-wine, but remember there were no "meds" in those days---that is the problem, the negative drug interactions. If you prefer wine, getting drunk is sin--through God's eyes your a hood. If you have to ask "After one, do I even think I could?" then you're probably *allergic*, which means *addiction*--craving the stuff, and it gets rough. For there's nothing worse than a hangover ("two four-letter words"): it's for the birds, this mountain of turds (that's not the life for the champion nerds). You must have your limit, and if you can't keep it you must (for the whole European times of day grape-therapeutic life) skip-it. If you can't sip it, enjoy diluted grape juice as the highest and greatest. The wine you must refuse: "just keep it". Now just enjoy the grape and your harvest (you will reap it).

ᚼCHRISTIAN HEDONISMᚼ

But: Wine or Welch's?

The Christian is supposed to enjoy life! Through the internet we've become aware of the supposed health benefits of grape (vino) therapy. However, let us reason together whether it should be wine or Welch's. All the new pro-wine literature says white grapes for the lung, red ones for the heart. In Italy people have always been so aware of "healthy wine" that they think it's unhealthy *not* to drink it. I always loathed wine for when young I drank with a wounded spirit--through me the devil did rear it and I ended up fearing it. The Roman and Jewish view of grape was "Christian Hedonism"--like good food, it's part of human life and should be enjoyed along with all other Godly gifts

171

JUST SKIP DINNER

bringing gladness to man. For years I took the Greek view which was to separate spirit from matter, also known as the the *yogi* route--*abstentionism* meant the bad was out. But now I see the Old Greek view is new age, not the "Christian thing to do". Now some of you will abstain because either you're happier on grape juice or take meds that miscombine. Because just about everyone takes some meds, this is the biggest reason for abstention--draw that line.

It's the grape getting you in shape. Biblically, no matter what we *want* to think, grape (wine) is seen as a gift from God and that was (and is) the early Christian view to more than a few. Calvin was paid with seven barrels of fermented grape, and the early Mormon apostles discussed scripture while drinking it. The Puritans had much wine, but then the non-biblical Greek view took hold and the whole Christian world did abstain--*legalistic mold dimmed the gold* of even the grape itself. The bible mentions grape constantly and there is no reason to think it had no alcohol content--it was more or less the same as now. Because of the Christian presence in Lebanon they are great wine-makers (wherever there are Christians there is great wine). But there are two ways of drinking (one good, the other bad) so if you're going to drink you must draw that line while always remembering it is PURE GRAPE JUICE for maximum health and shine:

⊰HEALTHY GRAPE NOT WINO-APE⊱

"One Glass for Females, Two for Men" But Some of You May Do Best on Pure Grape, My Friend

I have many Italian friends and they're going to drink wine--that's their only way to dine (it's like a shrine). I'm speaking to the others who may abstain but are thinking about starting due to the explosion of pro-wine sites who say to drink without refrain. Let's discuss the fermented grape and all the promised health benefits you'll supposedly gain: "One for women, two for men" was originally intended to restrict people from overdrinking but now it's become a health edict *against*

JUST SKIP DINNER

abstention, as if we should *all* drink moderate wine. It encourages moderation: one is not drinking to get drunk but rather for the *cumulative* health benefits from taking a little bit daily. If you look at the links at the bottom of this page you'll see these benefits pale compared to the destructive aspects of miscombining—people dying and we never hear about it. For example, wine *creates* tumors but pure grape juice *dissolves* them. If you are going to drink wine at least heed the warnings:

The Problem's Not Just the Wine—it's How It Miscombines. You could get roped into the Med Profession (separated from your savings) with just one miscombo

"But the early Christian mystics drank wine to get revelation". OK, I hear you. But since wine tends to dehydrate (it dries you out) one should drink it watered down (2/3 water) just as we do grape concentrate. If you're going to drink stop all medications or you'll be in trouble in a wink or tie up to a shrink. Older people often fall when drunk--that means a rest home with no more privacy (the beginning of the end all because one night they became a skunk). It's serious business, your whole life can change *suddenly* when you could've stayed independent, i.e. *heavenly.* The relaxing mild effects of a glass of grape is not bad to the Lord--getting drunk is the sin of the barbarian horde (and with meds combined you can be institutionally confined). They say 21 glasses a week, max. That doesn't mean all 21 glasses in one bender a week--but never more than three a day or you're up a creek (that's not relaxed). Now for lung problems, the white grape is said to increase in benefits with each glass. But of course after a certain line, you'd become an ass. Be mature--you'll know when you've too much sass. One of my readers said:

I have COPD and depend on albuterol in a nebulizer as my life-support. I was on numerous other medications. Well I started drinking white wine for my lungs and it miscombined--I could NOT get an in-breath! It was so frightening so I hit the nebulizer to open everything up and it got worse. I called 911 and they hauled me into a hospital where I stayed hooked up for two days to the cost of $5000 after which they wanted to put me in a hospice. I didn't think I could escape....what a nightmare, all because I wanted

173

JUST SKIP DINNER

to drink wine for the "health benefits"....One thing I remember is both the men in the ambulence and the doctors in the hospital denied that wine can miscombine! They are woefully uneducated about the problem, saying "it isn't the wine--get that through your head, it's good for you--we all drink it too". I will say this, without medications, it suits me fine. J. Kruse, your friend forever for speaking the truth

Alcohol and prescription or over-the-counter drugs can be deadly as four to 8000 Americans die yearly from the exaggerated (synergistic) effects which <u>suppress respiration and heart functions</u> to dangerous levels. It can also decrease a drugs effectiveness--suddenly it won't work "Alcohol and combining with other drugs,"
www.naples.navy.mil, 2002}

I tell you these pro-wine sites are creating destruction for countless millions who don't *respect* the wine by drawing the line. *Most* modern people are on meds over-the-counter or prescribed and with 150 of them there are really adverse reactions--even one aspirin causes stomach bleeding. A very common problem is acid reflux and these meds prevent the gut from breaking down acid *or* alcohol so one gets drunk faster. There are two kinds of drinking, one ok the other bad: The Northern Europeans had no wine with dinner but just got wildly drunk with gin and vodka (it was cold). But in Italy and France they drank moderate wine with long luxurious dinners and came out the winners: with low heart attack and alcoholism rate, devastation from alcohol was not their fate. Through the internet the cold climates are getting savvy about the wine-therapy with dinner and the appreciation of grape has never been higher: lovers of old grapes were *not* sinners but the Christian choir. Maintain thy shine by knowing thy line. And if you can't, abandon wine therapy before it turns unkind: Just take grape juice and you'll have beauty and shine. The grape is put on face in bath and every thing else--drink grape concentrate diluted with pure water for young elfs.

❧GRAPE CURE❧

JUST SKIP DINNER

I love pure grape concentrate in the morning diluted with water. All alcohol is a depressant so with someone who's *normally* high it can just make them cry. That was me many times--I wanted to die. The grape concentrate gives one the healing properties of grape without the fearful results of a *very* bad day. I take my grape juice in the morning then I fast until late lunch when I enjoy my peanut butter, sometimes a sliver of cheese, a little fish too--the Mediterranean Diet without the bread or starch, would this not please you too? I recommend pure grape juice for anyone seeking highest purity, with no alcohol at all--that's the fastarian champion crew.

Karen Maria: I'm so glad you're enjoying life now and I love your new direction. Let me say this: For years I too feared wine and smoked pot instead. I always felt it was far better since it doesn't lead to drunkenness, accidents, or jail. But now I see a big difference between that and vino or grape-therapy. I see what you mean now! We must see drunkenness as sin, but enjoy the times of day which grape is part of--grape is part of human life and always has been. The pot ruined my lungs, I have an embarrassing cough--that's as bad as drunkenness. I'm going to make the switch today and never smoke that stuff again. I'm going to start with merlot, the white for my lungs, get well, and just love my times of day life--esp the aft-fast blast and the luxurious nights preceding the happy mornings. I've enjoyed your website for six years now, you're never boring and I love how you change and grow. We all keep growing with you too! Mary S Johnston, Nebraska

❧THE FRENCH PARADOX❧

Most of the "health benefits of wine" literature is just the writer's self-justification to drink. They state the "French paradox" that despite high-fat diet there was low heart disease rate, due to the healthy wine. This has been disproven--the French Paradox is due to the *benefits of the high-fat diet* (*particularly* animal fat) not to the compensational qualities of the wine. High-fat is good, and *is* the paleo human diet for handsome human perfection from time's

175

beginning--the "compensation" of wine is often an excuse for sinning. Once one is addicted to drinking they will do and say anything--including leading the entire internet into sickness and death. Remember the entire frontal cortex of humans is there to self-justify, not "think" or "reason"--and nothing has this potential like alcohol (it's like treason).

Karen Maria--I joined the "Italian people meet" website and an Italian man called me "sweetheart"--I got sucked right in and he broke my heart. My God is there any hope for people like me?

KK: It's called the *child of the alcoholic (et. al.)* syndrome. Love-hunger interfaces perfectly with fast men. Watch out, then just go within. Solitude is your only hope, then when you find your self you'll attract your soul mate. To go "out there" without being whole is *just mean fate!*

⌐ΦEXCELLENT LINKSΦ⌐

Amidst *so much* false copy and self-justifying research comes these two gems which prove the opposite to all you've been told: On the problems with even frugal wine and how it miscombines: www.jrusselshealth.com. On the health benefits of high-fat (saturated animal) diet vs. the dangers of the low fat diet and other explosive myths and truths: www.westonaprice.org, especially the article "myths and truths" I strongly suggest you peruse these websites to balance the media's false pro-alcohol and low-fat reporting. For example, when advancing pro-wine studies the media always minimizes alcohol risks and maximizes benefits--or omits info allowing the public to see it as dangerous. The public deserves balanced and responsible reporting.

12

Fruit-Fat-Fastarianism

"Fruit and Fat is Where it's At"

Question: When and what do you eat and what's good about daily fasting? I love almond butter--is that good for the fat?

KK: What and when: Fruit smoothies, almond butter, grapes, Waldorf Salad made in the food processor with green apples, raisins, carrots and walnuts or low carb pizza made in the wafflizer with zucchini and tomatoes with cheese. I eat once a day after noon. The high-tech method is just skipping dinner for HGH (human growth hormone) comes out at night for HGH (if and only if) one is fasting and perfects (de-ages) the vessel no matter what the age. If every night you can just skip dinner you'll have a perfect life. Fat: Vegan dogma encourages nuts as the necessary fat but fauna is what we need for the reasons stated and just a little cheese with otherwise alkaline meals brings balance: "cheese based alkaline meals" is the title.

Question: Speak about allergy please.

KK: The worse your reaction the better you'll be without it. The more ugly you are, the more beautiful you'll be! If you've been allergic to a food you may have been extremely unattractive, lopsided or filled with unsightly protuberances like chronically bloated legs and ankles, puffy face, a mass of wrinkles. So if you give up that one food--for me it was starch and sugar--then you're whole life will change radically and you will look the opposite (normal, perhaps beautiful whereas you were quite homely). The worse you look (reaction) the more attractiveness without it (elimination) as indicated by your degree of sensitivity which created the reaction in the first place. Lives are changed with just this one detail--as the worst becomes the best (having passed the test).

Question: You talk so much about destiny. What is yours?

JUST SKIP DINNER

KK: In a world that never stops eating my destiny is to talk about fasting consciousness--how the fast is the feast vs. swine consciousness, where misery never ceased.

Question: What were you saying about a "mass of wrinkles"?

KK: At the base of every wrinkle is mucus and our mucus-level determines the "mass of wrinkles"--it's the same basis for one wrinkle or a mass of them. Women get depressed thinking their "loose flesh" problem is hopeless--it is not. Sometimes in just one day of fasting (after a lifetime of never missing a meal) the mucus level recedes and one becomes wrinkle-free over every inch of the body. The other reason for a "mass of wrinkles" is mal-assimilation. One may be skinny but there's still tissue-obstruction so new nutrients can't penetrate to perfect the tissue. The paradoxical answer for the incredibly skinny but wrinkled human is to FAST to increase assimilation so the tissue is perfected (wrinkle-free). To use a tragic example the emaciated concentration camp victims were nevertheless wrinkle-free. My parents fasted at the end of their lives and became wrinkle free whereas they were prune-like when middle aged. Often old age can be a beautifier as eating instincts are tamed and one conforms to these good rules.

Question: What about fat-fasting?

KK: What I love about fat-fasting is the appetite suppression, serenity and stability. For certain people with IR (insulin resistance) the fructose if eaten alone tends to create grogginess and craving. It's all about reversal dieting: whenever you hit a glitch, just switch. I love this life--you switch your hormones and you've switched your whole reality: (fantasiacal perception) of the properly fat-fed brain and fruit-cleansed bloodstream.

Question: How would you categorize your diet?

KK: It would be categorized thusly: Frugi-Fatarian-Fastarianism. Or: Reversal dieting between fruit, nuts and fauna {a little cheese]. Of you could call it "Bipolar (reversals) between High-Fat, High-Fructose and High Fasting Diet" This Reversal Fat-Frugivorous-Fastarian Diet could also be categorized as: "The Low in Starch, Vegetable and Flesh and High in Concentrated

JUST SKIP DINNER

Dried Fructose, Nuts and Cheese combined with Daily 20 Hours Fasting Diet". I think the name says it best: just FFF!

Question: What do you say about Atkins Dieters and also Ehretists?

KK: The Atkins Dieters eat too much, they eat too much meat, they eat too little fat (some actually buying low-fat meats still, hynotized by fat-phobia and despite how Atkins acclaimed it as essential and good), they eat too many meals, they don't fast enough, they eat too little fruit in fear of carbs and I doubt they'd ever eat raisins or figs--a "concentrated fruit sugar". The Ehretists eat too much fruit, don't fast enough, don't eat enough fat, have mood swings (sugar rushes and detox) and become increasingly dogmatic the more the diet fails them.

Question: Can you elaborate on food intolerance?

The bloated face and legs are one sign of food allergy--a reaction of a compromised immune system seen in many recovered anorexics. Food addiction, lowered immunity and then food allergies affected four systems: respiratory (lungs close in--difficulty breathing, wheezing), skin system (rashes-eczema-dryness-bloat), digestive (heartburn-vomiting-painful abs) and face—bloat, swelling, redness, contorted features. Due to low immunity I had these with fauna and nuts *at first* but with immuno-enhancement (reduction of total load combined with more protein) I was able to eat nuts or cheese with fabulous health. Just find the fat that works best and enhanced immunity will get those foods back--just eat that, reduce total load and enhance immunity and you'll regain a long list of foods with non-reaction. Universal reactors, unite-- just find the fat to bite. Most people love nuts and cheese because they *need* them and are *designed* to eat them--the fatty fiber in the nuts make them a superior food in three ways: low sugar, high fat and high fiber.

Question: I've often heard fats are constipating. What do you say?

KK: Nothing could be more mistaken. Although some fats like nuts or cheese cause allergic reactions in some people *at first*, the elimination is miraculously regular because the colon requires fat to "grease the runway"--the colon just requires fat, and that is that! You must actively resist silly slogans of the sick culture, for everything is the opposite to what we've been told! Cheese-phobics are reacting to their experiences when combining it with starch. And if grease brings on diarrhea, that is good! The evacuative functions depend on fat in the diet which becomes optimally efficient when combined with the high-fiber/fatty nuts. And tomavo, guacamole or Red Salad is very efficient without any allergic reactions for anyone bringing amazing daily evacuation like clockwork.

Question: What about sugar and mis-combining?

KK: The black fruit fuels the engine and cleanses the intestine. Enjoy your fruit sugar if there is no reaction, and the fat will bring balance along with the nuts which when eaten with the fruit slows insulin spike. Follow the plan herein and your problems with hyperinsulinism are over. Fruitarians overdo fruit-eating and don't do enough fasting. Even one apple or pear tends to effect me adversely when in the spike, waking up very groggy the next day. I may crave it but have to switch the desire from fruit to lemon-aid because fruit craving is often just thirst. *Miscombining*: I too was schooled to combine carefully—especially never to combine fats with sweet fruits--but now I see that as all wrong. I eat prunes or dates with cheese, and experience wonderful uplifting moods and evacuation each morning. Everything we've been taught is wrong—avoid dogmas of the throng.

Question: What physical changes should one expect as he goes through the transition to the fruit, fat and fasting miracle diet?

KK: The progression went something like this: After starting fruit-fat-fasting regimen the arms (lymph) broke out in a rash. After a few weeks I had much better definition in my (thinner) arms. After two months I started to *really* feel good which I think is half attributed to the diet and the other half to the daily fasting of 22 hours. After four months I started cleaning out deeper layers and all proportions were redistributed into a much more aristocratic posture—longer-thinner legs, more long-wasted, long and thin head shape and far more vigor than I've ever had. All wrinkles left the body as the problem receded. Simultaneously I was eliminating often as these layers unpeeled to the core. At the end I was thinner yet stronger than I've ever been having the shape of a young adolescent boy. (I mean long sinewy muscles--no middle-aged paunchiness or dowdiness). The skin took on a new luster and all age spots dissolved to reveal a ruddy complexion.

JUST SKIP DINNER

All wrinkles and paunches (jags and sags) from fruitarian water-retention left the face and it resembled porcelain. In other words the results of fruit-fat-fasting takes awhile but then you'll wear a smile. Try it you'll like it as you come into your own style. Now you can run a mile and be as sturdy, classy and clean as Italian tile.

I have seen these changes with absolutely everyone without exception: there is a conspicuous youthification with fasting--a de-aging of two decades with most. (Conversely one will "age overnight" if ever going back to starchy junk food.) Generally the face becomes much narrower and defines the individual—his facial features are more characterized as "himself"--the way he looked as a child. There is a much redder color over-all and clearer complexion. There will be a sudden diminishment of wrinkles and age spots all over the body. All muscles and tendon will be better defined as a young athlete. Yes hair often grows back and de-grays—depending on how genetic it is (and if it stays grey or white—it's beautifully distinguished like a dignified knight). Clothes will fit much better and stay cleaner (that's for sure--the pores are an elimination system for the CCC: constant crud coming through.). The body will lose all the ugly protuberances that define age. The "fortiesh" or "fiftiesh" look will go away to be replaced by the ageless glow ("have no idea how old she/he is").
The lips will be well-defined, the skin will have a real sheen from the fat and the eyes will shine like the stars at night having lost that dead hazy look of aging--impending death and despair. I often see these beautiful markers in people after they reverse to the FFF diet. Due to their renewed beauty they begin to love the fasting mornings the best when it is magnified.

Question: Karen why do I have so much problem not-eating, and why do I fear aging?

KK: You're always happy when you're eating. The biggest reason you eat is from fear, and by fasting you have to face your fears. Fear is the biggest reason for eating, and also boredom which comes from fear--since if we weren't afraid we'd be in our destiny, our own stream of creativity and constant delight. When we eat we bring the energy down from the brain to the gut--we can safely go into denial that way. So the question is: what in your life do you wish to deny? Let fasting reveal it so you can bid it good-bye. Mostly women eat at relationship problems and sexism in

181

JUST SKIP DINNER

everyday face-to-face encounters and most of that sexism manifests as ageism. Women automatically seek nurturance then, and eating is a handy way of denying painful reality. But fasting reveals the fatal flaws of seeking approval of society (of men mostly). Fastarian women replace cultural stereotypes for *Universal Archetypes.* The older woman has greatest dignity and most beauty--because it's something she's *achieved.* The older fastarian female is most cute with wisdom to boot. You'll love the diet as all aging fears are removed and you look forward to it all. As the years go by you walk more tall.

Question: Well then--how does one mini-fast?

KK: Ramadan Fasting is not eating during daylight hours, Buddha fasting is not eating after noon and mini-fasting is eating once a day only. Having two meals six hours apart (like breakfast and lunch) will give you 18 hours fasting daily which is great. Have your fruit in one meal and fat in the other if you want—I eat them together because I want the whole day to be in the fasting process. For the mildly clogged: As you fruit-clean you experience acute cleansing reactions like cold or flu symptoms, headaches or dizziness. But if you refuse to stop the cleansing with cold and flu medicines or wrong eating soon the process is complete and the cold won't drag on. On Monday sneezing chills and sore throat but by Wednesday morning there are only mild symptoms. The more clogged one is the more drastic and unusual the symptoms and the more magnificent after cleansing. Take it in stages like fasting week-ends followed by week-day meals. Nature heals in incremental stages--suddenly you see a much younger face in the mirror. Age is: *morbid accumulation.* Dissolve the chunk (the encumbrance) and you de-age.

Question: What is the main point of fruit-fasting?

KK: All foods except fruits and vegetables create acid or mucus-- an acid-mucoid plaque encases the intestines, arteries and every cell. The whole body from head to toe is a filthy storage pit of debris, a cess-pool. For me even cooked veggies created this debris as I bloated up like a couch—ouch! Fruit, Fat and Fasting brings streamline (that's the rare fin) as it cleans and burns this debris out--the basis of all disease, aging, ugliness and depression along with poverty. For if someone is alert and clear he finds his destiny and in a free country that means justified remuneration. To be at the top of our game we must be clean *and* properly fat-fueled.

Question: Can you elaborate on the different ways of daily fasting?

✑ RAMADAN FASTING ✑

I look at Ramadan Fasting as a wonderful "day-trip": the DAY is like a magic you step *into*: you explore the magic box all day until night falls, creating a miraculous right-brain attitude adjustment allowing a fabulous flawless day. I finish my work and breakfast before dawn, then walk much of the day in nature with my dog pack. I don't like eating lunch for it wrecks this trip: I want to have the food *behind* me so there are no thoughts cluttered with food. Just the mere intention to fast elicits PHF: Positive Healing Forces that flood every cell and make all wrongs "right", and it happens instantly. When I fast this way I'll eat peanut butter in the morning and a sliver of cheese at night.

✑ BUDDHA FASTING ✑

Buddha fasting is good for food compulsives. If you eat breakfast and are hungry later you can eat lunch by noon and still fast 18 hours. If one has eaten two meals, fasting by just skipping dinner is no big thing yet eighteen hours is a nice long fast as one feels a wonderful sense of accomplishment at night and the next morning. Done daily the results are cumulative, especially since one is sleeping in the fasting state when HGH is at its highest. In eating that noon meal remember that HGH is raised from protein more than any one thing except fasting. And so the paleo-faster is an ageless person. The most powerful Buddha fast would be eating brunch-only.

✑ MINI-FASTING ✑

There are some people for whom the first bite leads to an all-day binge. These types have found they must delay eating until the end of the day lest it turn dark on them. Mini-fasting is eating every 24 hours, usually dinner.   This is the most Ehretian--wait until 3 p.m. to eat the meal. The days are fabulous in the most ascended fasting state.

JUST SKIP DINNER

✀ FRUGAL DRIED FRUIT ✀

Question: You're big on dried fruit despite sugar content. Please explain

KK: With the dried black fruit like certified organic figs or raisons just a few are enough to assuage appetite for long periods. Never overdo the dried fruit which is highly concentrated but it's very handy for a desert or Eskimo recluse. With frugality there is no insulin spike problem, and because it is devoid of water there is no glycemic index problem of flooding the bloodstream with sugar, yet the wonderful vitarian properties remain—binding of mucus (twice as much as any juicy fruit) and debris in the stomach. The best part is: no matter where you are on earth you can have organic figs and raisins flown in so there's never an excuse not to eat fruit or take the grapecure. Anorexics stuck in hospitals should ask for permission to take the raisins/fig/nut with cheese route, for attempting to eat their prescribed food will end in pout. Tell the heads to call me—I'll let out a shout for logic and science does have clout (if you can get their attention, with all distractions out).

✀ CARBS-CANNED-CHEMICALS ✀

Question: What about all the carbs in sweet fruit and also your view on cooked food, canned fruit or chemicals?

KK *Canned or cooked food:* For certain problems canned or cooked food is more digestible: it's like *externalizing* the digestive process. For example many anorexics have mal-absorption syndrome (MS) so the best for them is (slightly cooked) canned pineapple chunks. Raw pineapple hurts my mouth and teeth but I eat a can a day of chunks in the hot summers—they are far more digestible than raw. It's wonderful to maintain a pantry and not worry about spoilage. Even naturalist Ehret ok'd dried and canned food in winter. Being a recluse who hates going to stores I load up on my favorites: canned pineapple in its own juice, black olives, fish, bottled lemon juice, unsalted tomato paste and 3-5 gallons of extra virgin Olive Oil. I don't like it when people bring me fresh fruit because then I'm forced to eat it lest it spoil. I can't stand having spoiling fruit around—for us fastarians canned fruit is best.

Carbs: Fruitarians consume hundreds of carbs daily and are still emaciated. The carbohydrate theory just doesn't apply to fruitarianism. Forget counting carbs or calories forever—just let the daily fasting regimen slim you down as it rights all wrongs--eat your full for the main meal. I'm not so much into lowcarb dieting but rather the high-fat and high-fasting fruitarian matrix.

Chemicals: Organic is all so expensive and usually you have to be in a city to get it. Isn't that strange? You'd think it would be easier out in the country, but not so. I've found a solution—buy certified organic dried figs (the fastest cleanser by far) through the internet and it keeps for months. That way you know you've had the best, free of all man's catastrophic chemicals for half of the day at least. However I have met many fruitarians who cleaned out on store-bought fruit. I'm not recommending it but organic is rarely available in a small desert town and a person has to adapt to the situation as it is. Fortunately organic *dried* fruit can be shipped in and you'll love your chemical-free fig and fun fantasiacal mornings. Since this life-style is about seclusion as much as eating right these internet tricks must go into the matrix.

❧ CELEBRITY HIGH-SPEEDS ❧

Question: Are there any celebrities you know of who fast?

KK: If Leo DiCaprio or John Travolta were to start fat-fasting they would regain the princely head shapes defining their early career. Age brings a box-head shape *if* one is eating wrong. "Prince" fasts a lot accounting for his great looks, often taking three days to disappear into his studio free of food and all human contact. (According to stories those working with him must fast too.) There was a time when Michael Jackson fasted every Monday beginning with orange juice but then relied more on plastic surgery and less on fasting. Plastic surgery is unnecessary if one is a daily faster for the planes of the face stay geometric and taut.

❧ ANOREXIA ❧

JUST SKIP DINNER

Question. What do you say to the Anorexic care-givers regarding diet?

Grains (breads, pastas, rice, cornmeal) are hardly cleansing foods. These foods fulfill most people's notion of "balance" but bring disaster to the ana when the significant others are planning her diet. Since many anorexic therapies refuse to let her plan her own diet these starchy mucus-creating dry foods ruin the immune system. They are too dry (only 10% water) and although they provide "fiber" are too concentrated and will not elevate consciousness to the high ecstatic state that fat-fruitarianism does—and only create an acid condition. The caregivers must allow her to stick to fruit and fat, then craving for bulimia will cease altogether. Now nuts and seeds are dry, but they are paleo so we've adapted to them, and they are also filled with fat and fiber (superior colon-properties). They are perfect for anas—tiny mouse-meals that nourish and sustain. But the other grains are agricultural—too new for adaptation and no good for anorexics with low immunity. Many immune-suppressed anorexics may be allergic to eggs and may also be sensitive about eating meat (that's me) so the next best way to get fauna is through cheese: lacto-fruitarianism combined with fasting. They may end up on some raisins, a few nuts and a little cheese for the day but recovered bulimics may have such suppressed immunity and food intolerances (increasing with age) that they end up on just lowcarb guacamole or Red Salad with non-sweet fruits with avocado. The more sensitive one is the more the suitable the cooked route, but it's surely not as "high" and that's what the ana wants--to be in the sky. She sought that goal through perversion—eating bad food to elevate morphine (created in the body with bad food) and then jettisoning superfluity to jet back up just as old sludge is pushed down. It doesn't work—she stays the clown. In these old failed fixes she'll only drown.

GRAPECURE

Question: What about the Grapecure?

KK: Energy really does come down to grape carbon just like Ehret says, so why not go right to it? The grape-cure also dissolves and pulls black grime from the intestines. It's the cleaning power that becomes quite evident the next morning. Even if the only grape one can find is Welch's grape juice he will still see black grime in

his elimination. It works! Grapes and later some fat--very French, very simple. All fruits translate to grape carbon but nothing works faster than the little ol' raisins and black fruit in general. As Berg's tables show the raisin cleans twice as fast as a grape and the dried fig is four times faster.

∽ EXERCISE ∽

Question: What about exercise?

KK: *Resistance bands and housework!* Fasting is more powerful than exercise. When one just eats fat and fruit the gut and intestines clean out and the abs become very well-defined. With this diet and fasting there's not that much need for exercise—but the body will *demand* it and you'll find yourself moving about: puttering (putting things into place) all day. I'm moving from 3 am until late at night, with only catnaps to recharge the system. But the FF-fed body desires it and can't stop moving nor can it sleep too long. The best exercises can be done at any age: walking and yoga. So much more can now be packed into each day that it doubles longevity. Through walking and yoga so much more is experienced in each moment, but if you want more I recommend resistance bands which snapped my body into shape quickly.

∽ WATER STATUES ∽

Question: What are your views on water?

KK: FFF resolves this problem and thirst leaves. Fat even creates more water--water is it's metabolic residue and fat brings water to the surface to moisten the skin. Most waters are bad anyway filled with chunks-called-minerals which turn water-drinkers into arthritic statues. Heavy water drinkers become become "dryer" each year (indicated by H20 in the body) due to hard bits in the water concretizing in tissue. I drink much lemonade and orange juice—a swig here and there rights the wrongs and everything is ok. For fasting use the "tinctured water system"—a couple drops in a quart. Get your water in fruit and lemonade and avoid all dry foods like grains, bread, rice and you'll rarely be thirsty. In fact the fastest fast is dry-fasting: eating or drinking nothing. Ehret never made a big deal out of drinking lots of water. All that water is a modern idea compensating for too much dead dry food. Of course you should drink as much water as you want if thirsty--but why do you think it's such a chore for most people to religiously drink 8 glasses of water a day? Because it's unnatural--drink only when thirsty. And when you desire juicy fruit, determine if you're just thirsty and if so continue the fast with lemonade.

✄ TOXIC PERSONALITIES ✄

Question: What do you mean by toxic personalities?

KK: It's a question of food and mood. There is a direct relationship and it's just about the whole determinant of mental state. The toxic personality was me. If I ate too much my skin became dry and I'd get "touchy". The system simply can't bog down or full-phobic discomfort takes over just as the fascination with miracles is lost. No more--I want that reality and have learned how to get and keep it: *rigid food routines.* If that makes me ana then I guess that's what I proudly am. I have pre-decided the best way to eat: fruits and fats followed by fasting every single day for life.

✄ IT HURTS GOOD ✄

Question: I tried the diet and got stomach pain. Shouldn't I stop?

KK: Celebrate all points of pain as proof that nature is doing her healing work. This is Ehretism. In fact Ehret recommends a three-day fast for the "magic mirror"--the first pain point indicating the area of one's major encumbrance. Take joy in pain and take the wider view--soon you will be through with a blissful future too.

Question: What about the permanent bowel dysfunction and mal-absorption syndrome in anorexics?

KK: The bowel loss will all even out with the Buddha or Ramadan fast. Eat the fruit and fat together and then not eating for 18 hours gives the elimination system time to work while everything miraculously comes out like clockwork the next day. This is always a pleasant surprise to most especially to anorexics who are characteristically constipated. And each day's fast improves absorption so that the body begins to assimilate food again--things don't come out just as they went in anymore in MS condition. In most anorexics the buddha fast alone improves all functions remarkably.

❧ SACRED SABBATH FAST ❧

Question: Can you elaborate on the "worthy week-ender", the Sacred Sabbath Fat-Fast?

KK: This is a time when we don't work and put away all petty business. This is to be a complete *reversal* from the rest of the week. Delete fructose for the week-end and just eat one big fatty meal like cheese omelet or doughless pizza to begin the day. This reversal diet from the all-fruit is a real boost to the system as it relieves possible constriction and has quite a slimming effect. I love this two-speed week when the week-ends are reserved for rest, reflection, real creativity. I usually start my fasting week-end on Wednesdays or Thursdays and it is the most productive time of the week though I always begin with the intention of rest and relaxation--that's how creativity works.

Question: Tell me more about Ramadan Fasting your lifestyle for life. Daylight hours are a magic box we step into? I love it but can't ever do it.

KK: One of these days it'll click. You will absolutely love the romanticism of it all. Yes it's like a magic box you step into, a fairyland filled with miracles to your benefit. Keep the daylight hours for miracles--feel like you're in a movie with you as star just for this DAY. If hungry just say "I'll eat tonight". If you can just get over the first hurdle--just for today--you're new life starts for the body instantly adjusts to the new habit. Additionally, fat-fasting heals dry skin quickly.

❧ FASTING CONSCIOUSNESS ❧

Question: What of the fasting consciousness you speak of so much?

KK: It's like one of my reader's said about the fleeting consciousness being "here there everywhere". She was amazed as fasting allowed a higher power to work everything out to perfection *through* her: *"The past gone, the Tao moves on, the future full-blown and nothing matters yet everything does: It's amazing how the mind wanders and jumps tasks very quickly when fasting. It's as if nothing matters yet everything matters so you don't waste time remembering or planning yet all you do is remember and plan. All of this is certainly a paradox. I am getting thinner and I love it but I like how the mind changes even more. The future isn't worth as much as the present and the past is a fleeting memory yet it all ties together and things don't matter but they do. A fasting day is a fascinating paradoxical day. It's another world that I never want to miss again."*

Question: Wow--this fasting consciousness is too much. It's like I do things but can't remember doing them.

KK: right--it's like a drug high. It is Fasting Consciousness as opposed to the omnipresent Swine Consciousness. You're going through the motions as spirit, not mind. Your actions are always right as you see in retrospect for a spirit doesn't think about what it's doing it just moves by Godly instinct. One should never trust his decisions while digesting. Once one sees all he's gaining by day-fasting there's no use to eating and with many it just starts a binge anyway. Unlike alcoholics we gotta have our fix once a day but anymore is addictive, unnecessary and often incendiary.
Don't forget the ana is "full-phobic"--even a little bit on her stomach and she's so miserable she says "oh what the heck....."
The early-only frugal fatty meal doesn't hit like lead--it just permeates right in. It catalyzes like a rocket-launcher while the second meal hits like a sledge hammer. It's like one meal is perfection and two meals may be the worst possible situation. Once you know about fasting consciousness you can easily substitute the highest reality for unnecessary eating.

✎ FASTING KEYS ✎

Question: Can you give us a trick to maintain the fast when things get rough?

KK: Let's say the afternoons are your problem point--that's when the food demons ordinarily hit. Since you haven't eaten since breakfast that's your bad time. Think of it this way: you now have *found time*--all the extra time (and money) from not engaging in that time-consuming and expensive activity. Now instead of binging afternoons you substitute with contemplation, yoga, meditation, study, taking a lovely nature walk or just looking out the window. For all that you used to lose with your food sin (money, friends, time, favor, beauty, destiny and God's rewards) you will now gain that much. Remember God's watching--and he's sure to reward your turnabout. Repentance is always rewarded with right-brain living where the magic is. All repentance is turning *from* the sin. Turn from, and the reward is much better than the prom. Just do it today and you will see--so in the aft' just take tea, or just a crumb.

✍ GETTING SKINNY ✍

Question: Why do people lose weight so fast on the Fruit Fast?

KK: Most of what we think is "fat" is just water retention and uneliminated feces from wrong eating. The average obese human has 25 lbs of retained water! Both leave quickly through the fruit, fat and fasting regime. It seems that instantly you're thin. The fat releases water from the system and the one fatty meal is an amazing laxative just as glucagons is elevated. So yes fat and fruit keeps you slim.

✍ PALEO ✍

Question: You speak so much about reversal dieting--can you elaborate?

KK: It's heaven on earth transcending the petty pains of food addiction and compulsive eating. A whole new world opens up with daily fasting after the high vibrational cleansing power of fruit combined with the pure power-punch and building from fauna-fat. Also fruitarians suffering with breathing problems will find them instantly relieved by taking in fat, which elevates glucagons which in turn dilates the airways (the opposite to popular opinion). Man has eaten that way from the beginning--fruit and fauna. The fauna-protein repairs so we can live basically on fruit the rest of the time without problems. The immune system is built back up through protein and fasting. This was proven to me when I became ill after breathing in wet paint for ten days. I became extremely weak, dizzy faint and deeply depressed. Instinctively I knew I had to eat protein and I instantly felt better and really came back to life. When things go wrong simply reverse. But always stay away from starch--the superior diet reversal is between fruit and (low carb) fat. You'll be so grateful you have something to reverse into when things go wrong.

✌ SICK CYCLES: BULIMIA ✌

Question: What are your views on bulimia?

KK: Bulimia is a Fatal Mental Disorder. Bulimics don't make it to old age, it's a lethal but very addictive habit. Anorexia creates emphysema by killing the lung tissue which gets oxygen. Isolation with a home nebulizer are just two results of this device of hypersensitives to "stay sane in an insane world". jettisoning all superfluity to garner control. A happy person doesn't have to do these things but refugees of dysfunctional systems and early trauma get hooked because Food Means Mother. It makes everything right.

Bulimia is now a world-wide epidemic. why? Concern for looks compensates for dysfunctional families. A childhood filled with coldness, rejection, broken family, addiction, alcoholism, violence, instability and incest is the basis of eating disorders of all kinds. Much of this assault on the soul happened at meal-time when families gather. The reactive craving for food becomes limitless like a possession by an outside force: you never know when it will "hit" and when it does you're helpless in it's grasp. Then the victim becomes "edacious"--eating ferociously like an animal. Many bulimics cannot recover until they see it as a sin. If they don't, it's always open to them--with stress, they fall into their bag or into death.

"To overcome such a thing brings unimaginable happiness"

It's like pulling through a war. But now every morning it feels so good just to be alive. The only answer to the food-trap is to FAST rather than FEAST and learn to love it more. If you used to binge each aft now glory in the AFT-FAST instead--it will make you so beautifully red and you won't be filled with dread. Choose the high-life instead.

Question: What do you say to bulimics?

KK: Most of your problems are due to this habit. You will only begin to understand the bad results later after your recovery. Even if people don't understand what your problem is they'll know something is "off" for it effects your whole personality and destiny. The consequences of this sin are inherent in the sin itself. You may think you're getting away with something--its a real "lark" at first but with time your life falls apart. It's a bad habit that discolors every facet of life. Be happy: replace it all with the fast and fruit life. The obsession will dissolve along with all fat and you'll be so proud of yourself for once. After twenty years of sobriety I have never relapsed and am only in disbelief it ever happened.

✎ FEMALE FASTERS ✎

Question: Karen, how does food and fasting relate to women's issues?

KK: Be a Nancy Reagan, not Rosanne. If a woman wants to maintain her dignity she should never be seen eating (or eating like a mouse). If a woman eats, she loses. To eat with people makes her one of them degrading her queenship.. The superior female must stay a saint--always fasting. That's how you win in both the public and the family and especially with men. Many men subconsciously pull back from women who eat while instinctively they know a fasting female is a winner--the elite. If you want to win at a normally losing game in a sexist society or system you must always put yourself in the position--through fasting--of looking down at the lusty lushes, the eating fools. Then you can love them despite their devices to keep you down. This may seem mean but it's actually a mental game we play in order to separate out from the demonic food life in which we were entrenched.

Question: Why do men hate to see a woman eating?

KK: Again, a female faster models herself on a Nancy Reagan archetype not a Rosanne Barr. Men hate to see a woman eating because it looks self-indulgent to them. Women are *archetypically* expected to serve men, not please themselves. When men turn against their wives for being fat it interfaces with the female's part in the interactional cycle of feeling rejected, and this makes her want to eat more for self-nurturance. This is a vicious cycle which ruins a marriage. Women, save your life and your marriage through the daily fast. Enjoy a good satisfying breakfast or lunch, then fast into bliss for the whole beautiful day. You'll be amazed how differently your mate responds to you (especially at night as you watch him enjoy the delicious dinner you have prepared for him but are above eating yourself). If you do eat with him (if I am taken to a restaurant, I eat salad and fish) eat very slowly in small bites and put your fork when you talk. No lady eats with her mouth full--what a shock!

✌ FASTING IN HOSTILE SYSTEMS ✌

Question: Does this not come out of your background in an anorexogenic system?

KK: That and my education in Systems theory. But through trial-and-error in the conspiratorial climate I learned all I needed to know in overcoming the hostile system. SYSTEMS THEORY will get a female victim into fasting quicker than anything--it was always these social aspects that did it for me. Kind of like priming as a movie actress or someone who's always in the spotlight (good or bad) and can only win by looking good all the time. In my family and marriage fasting appeased hostility while eating always increased it suddenly. That relationship is etched in my brain forever.

Question: Can you elaborate on the Systems set-up to which fasting is the only way to adapt, and win?

KK: Anorexics are always recounting their particular system set-up like the sister thing. She may be fearful in an identity struggle of two against one (triangulation-strangulation) and that template influences her life in a *hypervigilance* against triangulation— whenever she senses two against one, or just two *and* one, she's gone! It's a family system and if the victim-anorexic senses favoritism her identity is jeopardized and it feels like amputation. She eats "edaciously"--voraciously-- to assuage the awful pain. Binging is often an unconscious device to self-nurture in the face of this plight. If someone wonders why she can't lose weight, I'd look at that--feeling third-rate. The bulimic knows what she is doing but may not why. Then the situation worsens as everyone in the system *assumes* she's a vain egotist when actually it's a power struggle—that's the gist.

JUST SKIP DINNER

The victim may sense her parent's giddiness and changed attitude when the favored daughter comes around. Or she senses competition and jealousy coming from her mother. These Greek Tragedy nuances are actually *archetypes evoked in dense environments*. They are scary when played out and such a relief to overcome. I tell people to realize how much it hurts then test fasting as a way of dealing with it (while always remembering: "this too shall pass"). She will see that fasting is our trump card--it always wins when all else fails. Whenever there's a family crisis resulting in identity triangulation (interlocking jealousy patterns) try fasting as an divine opportunity to prove this out. It is the oldest way to win battles: Fast, and watch *them* eat cake.

✄ AGING AND FASTING ✄

Question: Can you elaborate more on how women—even as young as thirty--feel insulted about their age and looks and how it's resolved through fasting?

KK: An older woman has had her face pushed in the mud so much--by both men and women--she tends to "collect" these things in memory and by post-menopause becomes a prime candidate for fasting and the superior fruit-fat-fastarianism with no looking back. Like an anorexic she's found that food-control banishes ageism. Getting older brings rude remarks from Howard Sterns-types and their corny cronies but fasting stops up their mouths for now they're witnessing ageless beauty and allure, a universal archetype. This is a true lady—anyone insulting a true lady would be attacked by onlookers for you just don't hurt an obvious saint! Having achieved this new life aging becomes just a feather in your cap as you are actually getting more beautiful as the years pass, having learned how to fast, wash your face (book four) and handle stress. The older glamorous fastarian female is proof that "youthful beauty is an accident but older beauty is an achievement". It can be done--just fast, and go out in the sun. Both men and women adore the older woman who is ageless as proof they don't have to experience the dreaded decline. The whole race desperately needs models of "glamorous aging". This is one reason I love the people of India—they respect their elders and no one would ever think of insulting an aging female for they are the superior element protecting the hearth.

JUST SKIP DINNER

✑ FASTING WAR WITH FLESH ✑

Question: It's so hard for me to fast and eat only fruit. I'm young and haven't experienced these things. How do I make myself do it?

KK: You remind me of myself for twenty years. Every day I would intend to fast and every day I would eat instead. I came to realize it was a war between flesh and spirit. Being a highly self-willed person I could not fast for I did whatever I wanted. No matter how many resolutions I would make, on I would go--a mindless robot without control. Each and every day I would write in my journal "today, I must fast." And each day--particularly in the afternoons--the flesh and food demons got hungry and I would eat.

It was not until years later I realized it was the *fasting consciousness* that I loved--when I was ageless in the perception of beauty everywhere I looked. Only then did I have high self-esteem and complete optimism and faith. Once I realized that I never went back into all-day eating (*swine consciousness*) especially in a sick system holding me down (in which eating meant disaster). The trick is to make sure that your one meal is satisfying. If I ever did slip and eat I made sure it was with harmless and good things: raisins, nuts, a sliver of cheese, lemon-aid.

✑ THE BINGE ✑

Question: Can you elaborate on the binging of anorexics--those occasional "parties" of one?

KK: They are no fun though they always promise to be. That's the demonic lure--it looks like the cure. I came to realize that it never satisfied or satiated. It was just expensive in time and money. The mature ana comes to know that the only thing that makes her happy and satisfied is fasting. The binge is always followed by self-disgust--and this she can no more tolerate (it means rust). Only fasting brings true pleasure which is boundless and everywhere in the feeling of agelessness (no more fear). Like Buddha said, we "eat the universe"--and the rewards of austerity are great: you're now first-rate, with or without a mate.

JUST SKIP DINNER

✺ CHAOS ✺

Question: What about the chaos issue with anorexics?

KK: We easily feel it. For the cerebrotonic, the anorexic, the hyper-evolute ectomorph or the savant autistic, chaos is catastrophic. If things aren't in quiet order like a library or monastery the confusion is overwhelming. That's why disordered eating creates chaos in consciousness but it also creates compulsion to eat since it's "our time" so we don't feel the chaos of being imposed on by other people. But fasting calms the seas. That's why I recommend Ramadan Fasting, so we're never digesting in the presence of others (which makes the ana archetype impossible). Being geared for high creative action interactions bog the engine way down and there is a feeling of critical mass. We must live a separate reality or we are smothered. Anas suffer with fear or bipolarity (manic-depression) and fasting changes all this--even if ultra-thin, eating mouse meals of fat makes fasting safe and comfortable.

Solitude in nature or just your room: What a relief as the world falls away. Interactions on the path are often too much. Many cerebrotonics feared groups when entering Kindergarten and hated them ever since. Solitude soothes the sense of chaos ("burning building syndrome") that comes with all those emotions and energies. While fasting we are better able to handle interactions and stress--just eat right in the morning and let her fly. Fast, then trust your instincts and smile sweetly: "silent sensuality". It's true elegance and charm: she that says least is regarded as most intelligent while in the fasting state. "Women talk to much" men say--so be a lady: shut-up and just *intuit* today. It brings power, the power that comes from true humility-- knowing it's the fast, not us, that makes us great. We're just accessing God's divine device for man in times of trouble or vulnerability--which for cerebrotonic females may be most of their lives.

✺ SOLITUDE ✺

Question: You talk so much about solitude. Is this especially true for anorexics?

197

JUST SKIP DINNER

KK: It is not just an issue. It is THE issue, because

✑ "INTRUSION: WE CAN'T STAND IT" ✑

Anorexics can't stand intrusion of encroachment. That's what makes them different. Nancy Reagan once said intrusion actually caused her *physical* pain. Most people prefer a crowded beach to being alone in a desert cabin but the ana is so inward that loss of solitude is tantamount to torture.   An anorexic has certain limitations and people problems portend pandemonium. The privacy issue is so all-important that the loss of it degrades the physical. You can really see this in hospital settings which are often the worst possible situation with the domination of hierarchies and pecking-orders and enforced meal plans or "learning how to get along" with groups.   The mature ana has learned that staying healthy means staying fine, which means staying *free*.

The extreme ectomorph can love people, but still must be alone most of the time. Interactions don't just agitate they *aggravate*. The ectomorph's major system is the nerve and lymph system. The nerves wear thin under the skin and thus the desire for simplicity, order and quiet--going within. She may want music, but only her own. Beyond the touchy nervousness;   the ectomonrph is cerebrotonically in a *vertical* relationship to God not a *horizontal* one to her fellow man. That's just the way it is--she is separated unto God for a job. God designed the blueprint (opposing the mob). Success is soon--it's just a matter of being secure in the arms of He who provides the link to the Pot of Gold, and *knowing* success is sure.

Finding that Pot of Gold takes *solitude*: freedom for just pondering: right-brain, aimless thinking while looking out the window. It takes a lot of time alone to come into her own--that is, to realize her dream and give birth to God's vision for her. To the ana every day without solitude and fasting is a wasted day and every day with it becomes vitally important, magical and miraculous. She comes to success when she realizes that all she must to is fast [on food, people or habits] then ASK. Got it? I'll repeat: It's a matter of fasting, then asking.

JUST SKIP DINNER

Question: To sum up: please elaborate on the fruitarian movement man's protein and fat needs and can you express your views on dogfood?

KK: The mind and body needs ELIMINATION to come to the True Self. It isn't about having an encyclopedic understanding of the infinite varieties of plant-foods for man to eat. I already know my Creator is infinite but superior species are *oligiphagous*: *existing on fewest varieties for maximum adaptation.* All I want is my grapes, fruits, squashes, some cheese or nuts. This raw movement is still so much about eating: it's food preoccupation and addiction with a new image. What I'm into is *fatty potent meals then fasting for the whole day--the food is so potent that's all you need.* That's an entirely different life from gargantuan salads and all the fruit they eat all day and investigating all the multitudinous varieties of exotic superfoods and fruits all over the earth. Some sins won't go out but through fasting--but I don't hear the variety-obsessed talking too much about that as they age along with the rest. They put way too much reliance on food to heal rather than not-eating doing the healing (of both mind and body). Fasting every day is heaven. All this food-preoccupation is just a veil over our eyes keeping us from God.

Dogs and cats are dying of TUMORS as all the toxis in "pet food" accumulate. My. Cats are disgusted with their food from China. I finally just give them albacore, something real. And I feel sad for dogs of vegans--they don't get any fauna and fat which they need more than humans. Dogs love cheese--it's real food not vegetables or fruit—and they want fauna and fat first. Low-fat dogs end up fat with dry fur.

When my dogs showed dry skin and lackluster fur problems along with mites, worms, ticks, mange and other problems indicating low immunity I knew they needed a high-fat diet. With adequate fat they became gorgeous shiny puppies again. Dog diseases are rampant and vets are exorbitant. How ironic most vets are down on high-fat diets for dogs

✍ ANA-LITES ✍

Different Sides of Self by KK

✍ Chieftess ✍

I belonged in a zoo. Did food effect you that way too? I would salivate all day thinking of stew. I only wanted to know "how to make roux." That food created such mucus—so many colds, "ah-choo". But then I studied the Indian Way like the Sioux. I made fasting my life, no more looking like two.

✍ She Rap-Star ✍

I ate like a pig--I was as big as a rig, envying young girls lookin' like twig. Then I realized I didn't wanna be so big and that was all it took—I read Karen's book. Into my soul I took a look: was all that craving for love I mistook? Love God and myself--that's the Elf. When I put the food-fix on the shelf I became as swift and sure as the stealth. I can only say thank you God for such wealth!

✍ Desert-Lover ✍

I didn't needed a reason for lookin' like Jackie Gleason. Every bakery I passed the sweets called, they kept teasin'. I had no self-esteem—the demons were legion. God help me, to my identity this is treason! Then one day I learned true devotion to a much higher reality--it was like lotion as it set me in motion. I had so much to anticipate, like an ocean of miracles (of continuous notions). It's God's love-potion (as His child I got my portion). Eat and co-mingle, it's demotion. Fast in solitude: it's promotion. The best thing is: no more commotion.

✍ Den Mother ✍

JUST SKIP DINNER

All day long I dreamed about it: Food. It trapped me--this vice was lewd. I couldn't think of others: it made me rude and thank God no one ever saw me in the nude! But then one day I thought (I stewed): Can I ever be cute again, I cooed? Will I remain but a cow, I mooed? Then I knew: I'll just start to fast: *today* I've been cued.

✌ Thinker ✌

All I did was eat—it was no hard task. All day long I faked—I wore a mask. Then one day I decided to fast: in God's light I would bask. How did I do it? Of Him I had only to ask.

✌ Desert Queen ✌

It's turning to Spring--I can hear the birds sing! Before I was blind, and nature didn't mean a thing. But now that I'm a faster in love with The Master, my voice has a ring. I'm in God's favor—He's removed the food-sting.

✌ High-Ness ✌

I feel so stuffed, has my destiny been snuffed? In this chaos I've been roughed but now having fasted, I've been toughed. Bad food's been enoughed--with fruit and fat my hunger pains are muffed. Now my hair looks neat—no need to be coifed.

✌ Nature Girl ✌

I ate too much yesterday, so now there's hell to pay. I feel dumb with nothing to say and I'm not even excited about soakin' up the ray. All I wanna do is go to bed and pray. God said "but this too shall pass--just start a fast and this fatigue won't last". Soon it'll be out, lookin' like a pig's snout. I'll just endure and prepare like a good girl scout. Is there any other way to success, like a new route? No, this is it, so fast, rest and let out a shout!

✌ Sweet Lady ✌

JUST SKIP DINNER

I hate their guts, those demon-ruts. They make me such a klutz, just plain nuts. It's degrading, we thoroughbreds-turned-mutts. Well now I'm a fruitarian faster-- that's true guts. It's divine, its royalty like that of King Tut's.

❧Housekeeper: on Being a Pig ❧

I ate everything in sight--I am so contrite. It was never in God's light and it just wasn't right. God can you lend me your might? Food makes me mad, I am filled with spite! Delete my lax ways-- make my will tight so tonight I'll feel light.

❧ Me-Chief on the Past: ❧

I felt lousy all the time—I had no energy to go upwards nor to climb. All that crap I ate--what a crime! My insides were so filled with grime and my work wasn't worth a dime (one would never know I was in my prime). Whatever I tried to do turned to slime--I could hardly say it in a rhyme. But then when I learned to eat then fast on water and lime my life became great--thank you God for giving me the fast sublime!

❧ Susie Que on Joy ❧

I was a toad carrying a heavy load. It was a hard road but now I'm in a new mode. It's a serious code: eat then fast. That's a seed I have sowed and no more lines need be towed so just my own work has flowed. And then how I glowed—it's made me so happy in my abode, all blessings bestowed: so much joy I'm about to explode! All from turning down that pie a la mode.

❧ Me Ph.D. ❧

Even me Ph.D. had to pay a fee as the food-trap degraded my whole family tree. If only I didn't crave croissants with my tea! But then I started to really see—it was fasting consciousness that really pleased me and thee. Now I'm so happy--the fastarian's afloat on serenity sea.

JUST SKIP DINNER

❦ Arab Princess on New Days ❧

Sun, sand, moon and stars are so appreciated when the fast frees me from bars. If I eat early then fast I'm one of the Czars. I love my days relating to Mars. The fast killed the past and it removed all the scars. It dissolved all addiction--even my love of cigars. Now all I want is Spanish Guitars because we're Superstars in this New Life of Ours.

❦ She-Doctor on Fruit ❧

Embrace the fast more than the feast. If you're gonna eat keep to fruit and fat at least, for all else fattens like yeast. For the time of the fast it's God's bounty you've leased.

❦Dog pack Owner ❧

All those years I couldn't get enough. That's the nature of bad food--it's entrapping stuff, as my life itself was one big bluff. I kept asking "where is God?"--I couldn't take it on the cuff. Destiny was hidden--my work was a lot of fluff. Everyone around me was harsh (they seemed so gruff). Men, women, it didn't matter--they were all in a huff. But now as I fast I see it all new. Life is BLISS and I'm really tough.

❦ Ana on Thin ❧

I'm thin. It's in. But they tell me I'm a has-been. Should I take it on the chin? Well, I wear a grin 'cause I know fat's a sin (it's no way to win). Though they say "fat is fabulous" it's to their chagrin. I say go within--to your twin who is thin. That's the way to begin, therein. To get thin, get a thick skin--no more eating *at* them.

❦ Draw-the-line-no-Swine ❧

Ok I was a swine: I never knew where to draw the line. It was a demon force, that appetite of mine--I thought of nothing but where to dine. I never realized there'd be such a fine--a lack of shine. Then I'd medicate with wine (oh these habits of mine)! I'd bore my

203

friends with a constant whine, I just had no spine--gluttony was a sure sign. To God Dionysus my self-indulgence was a shrine as tall as a pine. I said "Can you forgive me God this wayward child of thine?" But then I learned to fast after fruit and fat to assign. My life became so happy, so *benign* when with God I aligned with proper foods combined. To hell on earth there was a sudden decline and instead I fell into God's design. It was DIVINE. My foes were all overcome--no more gossip to malign, my tastes refined and from the Devil's Crowd I did resign.

✧ Poetess on Bad Food: ✧

They called me "food-phobic" but now I see it was rational--it's eaters that are cracked. I had vision the other's lacked. I had a right to fear anything but fruit and fat--just have tact.

✧ Desert Doctor on New Life: ✧

I love this life. I eat then fast--its real freedom from strife. That old life of binging cut through like a knife. Now I just love resembling Barney Fife.

✧ Recluse on Igniting Your Dreams: ✧

Life was misery; it was strange like a grenade range. A constant bad-hair day--really a mange. Happy cordiality and friends I couldn't arrange. Everything I tried would always derange. From family and neighbors I would ever-estrange. From this to the good life I couldn't exchange. In business it was always me they'd shortchange. But then I met God and I saw the light. He said if I'd fast life would turn bright. All I must do is eat then stop--refusing a bite. Lord you were so right: the fast ended the blight! I started to win every time--no more *need* to fight. It was so wonderful, an all-day soul-flight (I felt as creative as Frank Lloyd Wright). I learned self-control with the strength of a knight. It took me to heaven--my life was out of sight. I became a little sprite with skin so tight and to those around me, no more spite. All day long it was constant excit, to my happy delight. So tonight I invite my friends to the daily fast so all their dreams will ignite.

✧ Me-Thinker on Treated Badly: ✧

JUST SKIP DINNER

I was so sad. You really treated me bad. You were such a cad I became badly clad--hardly starting a fad. Through you I'd really been had, so to hell with you my lad. You made me so mad but then I went to God my Dad. Now I'm so glad and after this war-story I'm a grad. Now I make a castle out of my pad. A faster and thin, I can even wear plaid.

❧ Me-KK on Treat Me Right ❧

How could you be so mean? I was only a teen. Well now I've got lean. With the fast I glow--I'm in the sheen. That's a True Leader, I'm talkin' Queen. To get there I just turned down the bean--I got really clean and even got the attraction of the Dean. It's a genius-gene making me so keen, and I'm the best-lookin' lass you've ever seen. It's my new scene being ready for the screen. I'm gonna be a star with no-loss in between nor need for caffeine. To look eighteen, just refuse cuisine. Then tough as marine, accurate as a machine--that's just routine. My genius lays unseen. But by eating cheese and tangerine it's never obscene and so serene.

❧ Indian Lady ❧

Since I fast I'm in a trance: miracles way beyond mere chance. I'm so energetic I just have to dance: my mind is everywhere, always--viva la France! I'm surrounded by spectators giving more than a glance. I'm filled with Holy Spirit--a very strong stance. God's child is a millionaire--He's given me an advance. My skills now so exquisite--the fast has enhanced. I love the daily fast--I'm wild with Romance!

❧ View-Lover ❧

On my shoulder a giant chip. Into multi-addictions I would dip. They called me boring--I was a drip. Yet with the slightest irritation I would flip. By all the food demons I was in a grip. To my generation I was hardly hip. The younger kids they'd give me the lip, and with business relations I'd get the gyp then of wine and beer I'd take a nip. So I thought I'd write to Karen in this quip. Then I gave the fast a rip and I won't jump ship. I ate breakfast then stopped--just lime-aid, a sip. Within a few days wow! I'd not

205

walk but skip. Off of this wagon I shall not slip--what a way for the Saints to equip! I've never had so much zi—it's the way to give the devil a whip. Let me give you a tip--become a happy Hopi faster then be proud as you strip.

↬ Ponderer ↫

I've always sought truth. At fairs or the circus I'd seek the science-booth. I had a mind of Einstein--I was a sleuth. Then one day I got old--long in the tooth. I didn't like the look (I found it uncouth). In shame I sought Vermouth but that was not the Fountain of Youth. So I started to fast and regained my youth--at last, the Truth.

↬ Marikelok ↫

All-day eating made me a sleaze. I was a glutton, not busy like bees. Self-esteem was so low I just wanted to please. I was often sick with coughs or an embarrassing sneeze. In public places I felt so ill-at-ease: around people I got the big freeze. This eating thing was really a disease. I had ups and downs--it was like a trapeze. I turned nothing down anything with cheese. Since it was mixed with starch it added bloat--it made me wheeze. I got fat in degrees--it was me in threes.

Finally I asked myself: Gee, Louise. Don't you want a fresh breeze? So what if the world disagrees, just fast--the results God guarantees. I ate once then fasted, I talked to other food-escapees. I met God and became a fast-devotee. After one week I was in bliss--I'm one more that agrees. I'm gonna fast every day--I'm a life-enlistee. Another way of saying: I'm a gut-amputee.

My work, my hobbies really showed expertise. My face took a glow, my eyes swooped up like a Japanese. Having eaten I was able to say "NO" to food--the devil was displeased. As I aged I got better--a heaven appointee. What a mind-trip: such beauty and bliss--like exploring under seas. My body got taut without exercise--it attracted a squeeze: I walked over hills and valleys easy with new legs such as these. The bakery had no power--it had lost its tease. Whereas skin was brittle and dry (like living with fleas) it was now so moist--I thanked God on my knees. They said "How did you do it, your new life of ease?" I'll tell you by giving you the keys: Eat fat and fruit from the trees then fast. Your youth will return: just do it and see as all problems cease.

⚜ KK Food Bio: ⚜
Daily Fastarian Solitude

The Fat-Fruit-Fastarian Diet is Fruit
or Fat once (best) or twice a day (OK)

HIGH FAT Question: "Won't that raise my cholesterol?" No the high fat diet will lower it but you must abandon all refined starches and sugars so your system can heal, clean, rebuild and become perfect (youthify).

FRUIT Question: "sugar brings insulin spike but fructose is released slowly into the bloodstream right?" Juicy fruit (juice) releases into the blood very fast and if there's a spike you'll get hungry. If no spike, great—enjoy. Black fruit cleans fastest and raisins and figs are far more potent in vitamins and antioxidants and won't bloat you up--a little goes a long way just like the potent fat.

FAST Question: Don't I need a long fast? "No daily mini-fasting is sufficient to heal your body and restore your youth. Eliminate all starches and drink coffee or lemon aid.".

Sample Meal Plan: raisins (figs prunes) when first hungry Red Salad or doughless pizza or cheese omelet then with 18-24 hrs fasting a day

⚜ FRUGAL FRUIGI-FATARIAN ⚜
FASTARIANISM:

JUST SKIP DINNER

I'm so happy now. A few nuts, raisins or cheese then I fast for the day. I sleep so well the minute I hit the hay. It's so romantic the fasting consciousness--it's more than ok. It's so deep my ascent so steep I'm rid of that creep I feel so much peace (from food demons not a peep). You gotta know you're fasting--that's the bliss. To feel so proud and to enjoy your day as God's kiss. He rewards the faster with nothing amiss. Everything forgiven, enemies overridden and sure success soon--that's a given. I'm not kiddin'--do it today you'll be believin'.

Karen Kellock
Struggles of rising back up again.
(Lowfat is not the way to thin).
1979

1979 Paleo Diet: Paleo diet and temporary success followed by the fall into first-fame syndrome: the "dry silent years of many tears".

1995 Fruitarian: Ten years out of society. Tomavo in the morning followed by walking and biking in 120 heat. Lack of fauna-fat ends in failure to thrive and shaky isolation.

1999 Fruitarianism near-dead. Non-fat fruit diet ends in hospitalization emaciation, fatigue, breathing problems and environmental illness (extreme immuno-suppression)

2000 Frugi-Fat Fastarianism: Emaciation gone and health restored with fat, fruit and fasting.

Fruit, Fat and Fasting is the heavenly triangle. Reversal dieting between Atkins (fat) and Ehret (fruit) is the answer you've been looking for. Bliss is not the Riviera or Vegas but your own home in the fasting state in which a miraculous new life opens through an inner journey. Daily fasting is the key! And higher paleo fasting is based on the essentiality of fats which put the body in fat-burning mode bringing energy and appetite suppression for the daily fast. It's a blast! From fats reverse into fruit fasts like the grapecure— for energy comes from grape carbon as well as glucagons-elevation from fats.

JUST SKIP DINNER

Karen Kellock Ph.D.: The answer to all our miseries lies in the reversal of diets. It's not a controversy between fruit or fat—between Ehret and Atkins—but rather becoming a very wise person due to our knowledge of both and *knowing when to reverse*. The joys of daily fasting are boundless and cumulative—every day you're cleaner, happier, younger. The Just Skip Dinner Club is pure class: the aft-fast is the Glory Church. Congratulations—today you've found the key and it's all for free!

৬৯৬৯

✍ BIBLIOGRAPHY ✍

Aiello and Wheeler, in *Current Anthropology,* 1995.

Atkins, Robert. *Dr. Atkins New Diet Revolution,*
Avon Books, 1992

Benoit, F.et.al. "Changes in body composition during
Weight reduction in obesity," *Arch. Of internal
Medicine* 63:4 (1965), p 604-612

Billings, Tom. *Beyondveg.com*

Cassidy, Claire Ph.D. in *protein power,* p 401

Cockburn, *Mummies, Diseases and Ancient Cultures;*
In Eades, p. 400

Eades, Michael and Mary Dan M.D. *Protein Power*
1996, Bantam Books

Howard, Vernon. *No, 50 Ways to Escape Cruel People,* 1981

James, William. *Varieties of Religious Experience*, 1858.

Lovewisdom, John. *Vitarianism,* unpub. Manuscript, 1975.

O'Dea, Dr. Kerin, in *Protein Power,* by Eades and Eades p. 46

Schachter, Zalman and Ronald Miller. *From Age-ing to Sage-ing*,
Warner Books 1995.

✍✍✍

100 KAREN KELLOCK BOOKS

AFFINITY OR MISERY
AGELESS CORNUCOPIA
AMERICA AWAKE!
AMERICA'S DAFT ERA
ARTS OF PALEO FASTING
AUTOPHAGY ON CHEATERS
BACKSTABBING NEUROTICS
BETRAYAL TRAUMA
BOOMERS AND BROKENNESS
BOOT ON NECK
CHAMPION GUIDES
COMMIE NUTHOUSE
COMMIES
COMMUNIST SPIRIT
CONTAGION OF MADNESS
CONTAGIOUS MADNESS
CULTURE CLASH BASHED
DAFT LEFT
DAILY FASTARIAN
DAM RATS
DIVERSITY IS CRUELTY
E-RACE WHITE
EVIL FREAKS (Beyond Gross)
THE END OR A BEND?
FEMALE BULLIES AND FEMI-NAZIS
FEMALE CARNALITY
FEMALE DUMB DOWN
FEMALE POWER DRIVE
FEMINISM AND RUIN 1 & 2
FIX FOR MISFITS
FOOLS & TRAMPS
FREEDOM SPEAKING
FRENEMY ENABLER
FRENEMY LIAR
FRENEMY THIEF
FRENEMY TRAITOR
TRENEMY TYRANT
GENIUS IS HELD DOWN
GLOBALISLAM
GOD USES THE FLAWED
HAZE OF THE LATTER DAYS

AUTHOR BIO

Karen Kellock Ph.D.

Ph.D Political Psychology, UCI 1976
Post-Doctoral: UCI Medical School
Department of Psychiatry
Grants NIMH, NIAAA

Ph.D. dissertation "A Systems-Theoretic View of Pathologic Interaction" made an early mark as the "Wife of the Alcoholic Syndrome". Postdoctoral research at UCI Medical, Dept. of Psychiatry on the systems surrounding pathology on NIMH and NIAAA federal grants: *The Contagion of Madness: The Psychology of Neurotic Interaction and Pathological Systems*. Therapy tool Therapeutic Playwriting introduced the play *Mary and Murv: Gruesome Twosomes in the Alcoholic Marriage*. She taught Abnormal Psychology and Pathological Systems Theory at UC and CSU campuses and developed "the Debris Theory of Disease" in five books and website: (www.karenkellock.org): *Champion Guides, Daily Fastarian, Just Skip Dinner, Arts of Paleo Fasting, Ageless Cornucopia. Manual for Superior Men is a* pick-it-up-anywhere book that you can't put down (20,000 Kellockialisms) and ever on your desktop it should be found (or this Ebook for superior wordsearch of new jargon).

www.ingramcontent.com/pod-product-compliance
Lightning Source LLC
Chambersburg PA
CDIIW071221290326
41931CB00037B/1754